CONTENTS

Figures and Tables	3
Foreword	5
Yurruungga Aborighinal Corporation	7
About the Friends	8
Preface	9
Introduction	13
Conservation Values	17
Forests	17
Koalas	21
Summary of World Heritage Values	25
Findings	26
Recommendations	27
Friends Reserve Proposals	31
Friends of Coffs Coast Koalas (FoCCK)	32
Friends of Conglomerate (FoC)	44
Friends of Dorrigo Koalas (FoDK)	56
Friends of Kalang Headwaters (FoKH)	68
Friends of Newry (FoN)	80
Friends of Orara (FoO)	90
Friends of Tarkeeth Koalas (FoTK)	104
Friends of Tuckers Nob (FoTN)	114
Further Reading	127

Tuckers Nob eucalypt forest

FIGURES AND TABLES

TABLE OF FIGURES

Fig. 1	Greater Koala Park: Tenure	10
Fig. 2	Greater Koala Park: High Conservation Values	18
Fig. 3	Greater Koala Park: Koala Species Sightings	22
Fig. 4	FoCCK - Proposed reserve	36
Fig. 5	FoC - Proposed reserve	48
Fig. 6	FoDK - Proposed reserve	60
Fig. 7	FoKH - Proposed reserve	72
Fig. 8	FoN - Proposed reserve	84
Fig. 9	FoO - Proposed reserve	96
Fig. 10	FoTK - Proposed reserve	108
Fig. 11	FoTN - Proposed reserve	118

LIST OF TABLES

Table 1.	Criteria for assessment and indicative justification for listing	25
Table 2.	Parkwide data	28
Table 3.	Friends Reserves data	29
Table 4.	Other Proposal data	29
Table 5a.	FoCCK - Reserve data	35
Table 5b.	FoCCK - Koala food trees present and regional use rate	38
Table 5c.	FoCCK - Threatened species	42
Table 6a.	FoC - Reserve data	47
Table 6b.	FoC - Koala food trees present and regional use rate	50
Table 6c.	FoC - Threatened species	54
Table 7a.	FoDK - Reserve data	59
Table 7b.	FoDK - Koala food trees present and regional use rate	62
Table 7c.	FoDK - Threatened species	66
Table 8a.	FoKH - Reserve data	71
Table 8b.	FoKH - Koala food trees present and regional use rate	74
Table 8c.	FoKH - Threatened species	77
Table 9a.	FoN - Reserve data	83
Table 9b.	FoN - Koala food trees present and regional use rate	86
Table 9c.	FoN - Threatened species	89
Table 10a.	FoO - Reserve data	95
Table 10b.	FoO - Koala food trees present and regional use rate	98
Table 10c.	FoO - Threatened species	102
Table 11a.	FoTK - Reserve data	107
Table 11b.	FoTK - Koala food trees present and regional use rate	110
Table 11c.	FoTK - Threatened species	113
Table 12a.	FoTN - Reserve data	117
Table 12b.	FoTN - Koala food trees present and regional use rate	120
Table 12c.	FoTN - Threatened species	122

The coastal forests of the Mid North Coast of New South Wales provide some of the best remaining contiguous stretches of Koala forest in the country

Coastal eucalypt forest

FOREWORD

Koalas are an international icon for conservation species, rivalling the panda and polar bear for global recognition and concern for their future. Yet the fate of Koalas is inextricably linked to the unique eucalypt forests to which they are adapted. Koalas are not only dependent on the highly toxic eucalypts for food, but they are symbiotically adapted to feed only on a very specific range of eucalypts, which may only be edible at certain times and in certain places. Only a Koala can tell you what habitat and what food it needs at any given time. But the one thing they all agree on, is that they need a very large area of forest in order to live, breed and survive into the future.

The coastal forests of the Mid North Coast of New South Wales simultaneously provide some of the best remaining contiguous stretches of Koala forest in the country, a little over one third of which of which is formally protected, and are steadily vanishing due to ongoing land clearance through forestry, agriculture and development. These forests lie at the intersection of Australian northern and southern forest bioregions, creating a region of high floristic and faunal biodiversity and endemicity. Despite years of effort to protect them, these spectacularly beautiful and diverse Australian forests continue to be logged, cleared, damaged and fragmented.

Protecting the remaining Koala forest habitat on the east coast is an absolute priority for Koala conservation nationally. Koalas feed on over 70 species of eucalypts but any one Koala, in any one area, can only digest the leaves of 4-10 tree species. Of these, palatability varies by tree age, soil type and water availability and between different leaves. In addition, Koalas require a range of heavy foliaged shelter trees to protect them from heat and predators. Only by protecting a large and diversely structured area of natural forest with a mosaic of forest types and ages can we ensure adequate food and shelter for the survival of Koalas. The highly specialised diet of Koalas, and the unpredictably erratic nature of Australia's drought cycle, means that Koalas require vast areas of habitat to survive. Koala home ranges vary from one hectare (the size of an average sports field) in the most productive forests to as much as 300 hectares (the size of New York's Central Park) in drier forests. In captivity, each Koala requires a plantation of over 200-400 specially selected feed trees, making them one of the most expensive captive animals in the world.

Walking a metabolic tightrope between toxic food, drought and fire, the Koalas and their forests currently face ever increasing threats from development, land clearance, disease and climate change. The Black Summer fires, which saw the deaths of an estimated 45,000 Koalas, illustrated just how vulnerable this species is in our rapidly changing climate. Even in their heartland, in the coastal forests of northern New South Wales, Koala populations are declining and action is urgently needed to protect their remaining habitat. Koala were once one of Australia's most widespread species, occurring in forests from far-north Queensland down the east coast, throughout Victoria and into South Australia. Land clearance and hunting brought them to near extinction in the southern parts of their range by the 1930s and only a concerted international campaign brough the export of millions of Koala pelts from the forests of New South Wales and Queensland to a halt. Since then, Koala populations have recovered in some areas but continue to decline across the eastern forests that were once their heartland, retreating into increasingly isolated and shrinking pockets of remnant forests, surrounded by inhospitable and hazardous roads, houses and farmland.

Koalas are not only a flagship species for international conservation, but for the Australian eucalypt forests they depend upon. Protecting their forests under World Heritage status, not only protects this iconic, internationally loved species. It also protects rich and complex ecologies of ancient Gondwana, rainforests, modern sclerophyll forests from the subtropical, warm and temperate zones, along with the rich abundance of rare and remarkable species that live within them.

Danielle Clode
Author, Koala: A Life in Trees
Associate Professor, Flinders University
Member, Koala History & Sustainability Research Cluster

Dean Thomas Kelly, CEO Yurruungga Aboriginal Corporation

YURRUUNGA ABORIGINAL CORPORATION

YURRUUNGGA

As the CEO of Yurruungga Aboriginal Corporation, I am proud to express our unwavering support for the Greater Koala Park World Heritage Proposal. Our dedication to the conservation of our cultural and natural heritage resonates with the goals of this proposal.

The Greater Koala Park is not just a haven for the koala populations but also a symbol of the cultural richness of the Gumbaynggirr Country. The ancient woodlands within the proposed park are integral to our history and traditions. They are sacred places where the Gumbaynggirr people, known as 'the sharing people ', for the generosity of our lands, have prospered for millennia.

We acknowledge the pressing need to protect these landscapes from the impacts of climate change, deforestation, and other environmental challenges. The World Heritage designation would protect the koalas' habitat and ensure the conservation of our cultural heritage sites throughout this landscape.

As the traditional Custodians of the Bellingen LGA area and the Budaabang Clan of the Gumbaynggirr Nation, we emphasise that our guardianship extends exclusively over the Bellingen valley. It is crucial to note that the Friends' groups associated with this proposal do not represent Yurruungga Aboriginal Corporation.

Our support for this proposal is based on our commitment to protecting the forests within the Bellingen valley, which aligns with our focus on preserving the environment and our cultural legacy. We are eager to collaborate with all stakeholders to achieve a successful World Heritage listing. This initiative will enable us to apply our traditional ecological knowledge, promote cultural education, and share the significance of these lands globally.

We are ready to support the implementation of this proposal, which we believe will be a significant step in recognising the unceded sovereignty of the Gumbaynggirr people and the enduring connection between our culture and the land.

In solidarity with the spirit of the proposal, we anticipate the positive impact that the Greater Koala Park World Heritage listing will have on our environment, our culture, and the generations to come.

Kind regards,

Dean Thomas Kelly
CEO
Representing Yurruungga Aboriginal Corporation

Aniseed Myrtle (Ringwood), Tuckers Nob

ABOUT THE FRIENDS

The Friends groups comprise local residents who live in and around the State Forests and National Parks under consideration for inclusion within a Koala park for the Mid North Coast of New South Wales.

As members of the community, they come from all walks of life, political persuasions and cultural backgrounds, but hold one thing in common: a love of the particular forest they call home, and the plants and animals they share this with, especially the Koala. They have arisen in recent years in the wake of escalating climate change and habitat loss to volunteer their time and advocate, using citizen science. They support the protection of public land on behalf of all forest dwellers, human or otherwise, to ensure all can survive and flourish in an environment of natural and cultural integrity.

The Friends acknowledge the traditional custodians of this Country as the Gumbaynggirr people and recognize their continuing connection to these forests, waterways and wildlife.
We pay respect to their Elders, past and present and their descendants.

PREFACE

6/05/2024

From: Koala History and Sustainability Research Cluster

To: Friends Groups and Readers

Dear Friends Groups and Readers,

I am writing on behalf of the Koala History and Sustainability Research Cluster (KHSRC; https://www.khsrc.com.au) to advise you that we have read and examined the *Australian Eucalyptus Sequence World Heritage Area Proposal, Stage One: Greater Koala Park* ('Greater Koala Park World Heritage Proposal'). We are happy to be associated with this project and endorse its contents. KHSRC is willing to continue its involvement with this initiative as it progresses and to provide input and advice when the State and Federal Governments prepare and submit their nomination for World Heritage listing.

KHSRC commends the Friends for their hard work in initiating this bold and timely vision. If implemented, World Heritage status would mark a big step forward in protecting one of Australia's most important populations of wild koalas, its hallmark *Eucalyptus* forests and the many and varied sequences in which they find their expression. May this proposal motivate not only the Federal and State Governments of Australia, but the other countries of the region that are home to this most remarkable of trees, and where the flagship species Koala is endeared.

Humanity stands on the precipice of change. Science tells us that many of the Earth's systems are under severe strain, and the majority of its planetary boundaries have been exceeded. We need our leaders to take action to combat climate change, biodiversity loss, deforestation and desertification. That requires aspiration, and aspiration requires courage and meaningful effort. The work of the Friends, based on sound research, commitment to community, and care for Country, offers a promising way forward, which KHRSC hopes will encourage many others to follow.

Yours sincerely

Dr Rolf Schlagloth, Founding Member, KHSRC

For

Dr Flavia Santamaria
Dr Alexandra McEwan
Dr Gabrielle McGinnis
Prof. Fred Cahir
Dr Tim Cadman
Dr Michael Danaher
A/Prof. Danielle Clode

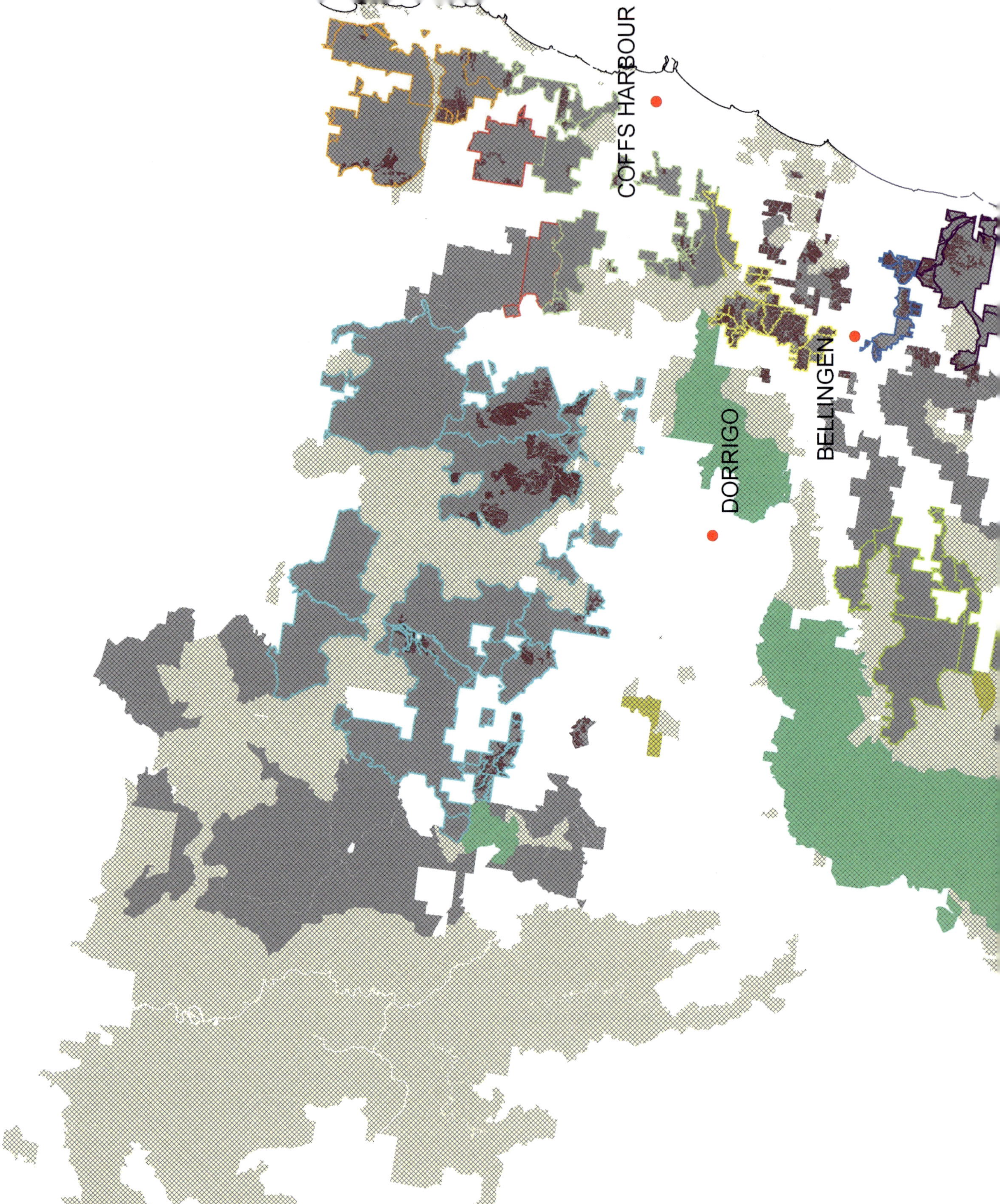
COFFS HARBOUR
BELLINGEN
DORRIGO

FIG. 1

Greater Koala Park:

TENURE

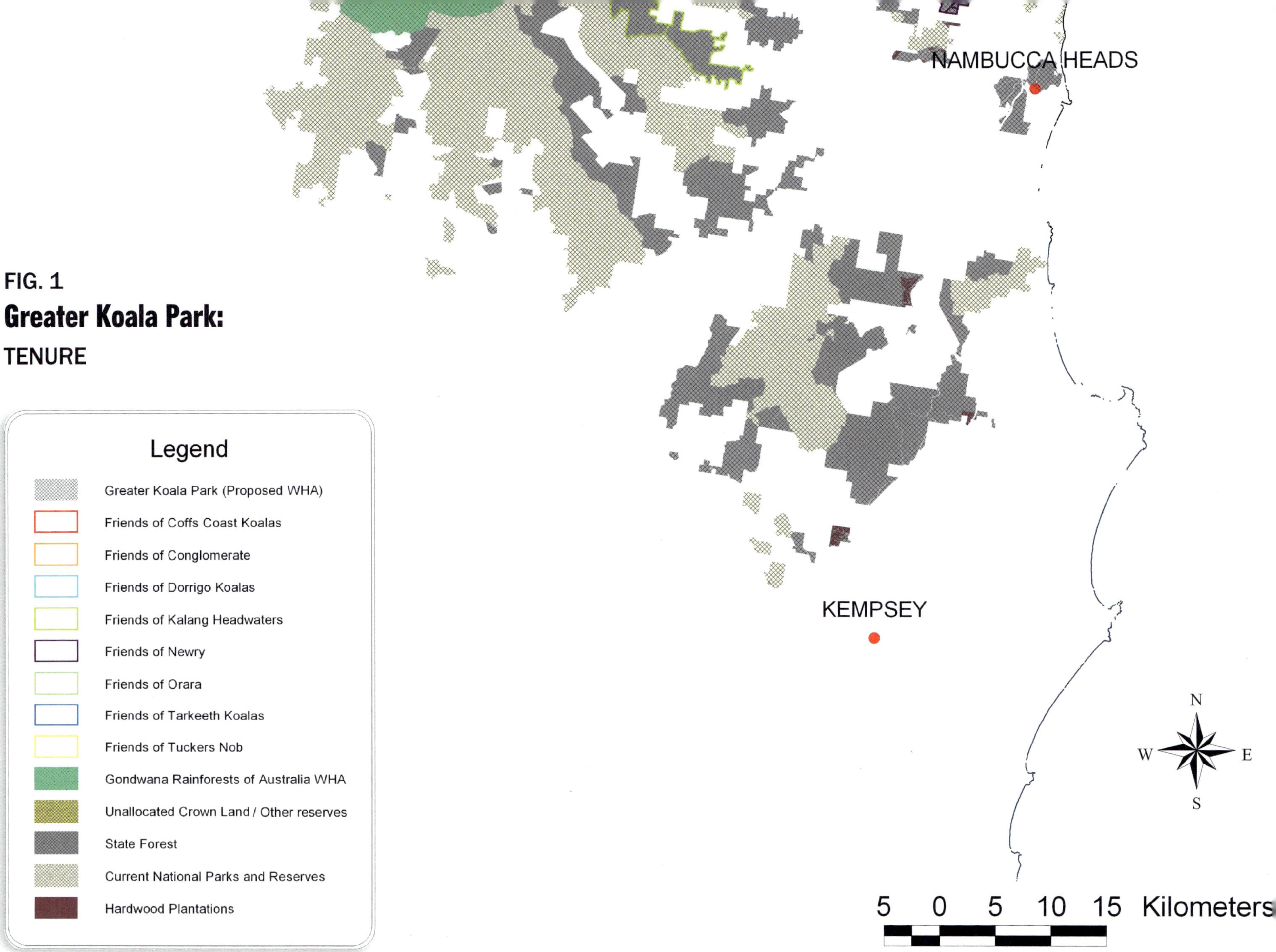

Ancient giant, Nana Creek

INTRODUCTION

BACKGROUND

This proposal for World Heritage listing of the *Eucalyptus* forests of Australia, and the Koala habitat within them, comes at a critical point in the Earth's evolutionary history. For the first time, the impact of one species, *Homo sapiens*, is having impacts at a global scale, affecting almost all of Earth's planetary systems, from the biosphere, to the atmosphere and beyond. In Australia the aftermath of the Black Summer bushfires of 2019-2020 has left many of the affected forests of the East Coast exposed to ongoing weed invasion and soil erosion; some are on the brink of ecosystem collapse. This was particularly the case for the forests of New South Wales, which witnessed the loss of many precious rainforests dating back tens of millions of years, to the ancient continent of Gondwana. Arboreal mammals, especially the Koala, suffered heavy losses of local populations, and it is this, in part, which led to the Federal Government upgrading the conservation status of this beloved national and global icon from threatened to endangered. In 2023 the new state government recommitted itself to protecting the Koala within a reserve on the Mid North Coast of sufficient size to ensure the protection of this endearing species and its habitat. That process is now underway and local communities and scientific experts have rallied behind this cause. They are offering this proposal for a Greater Koala Park as a statement of their own commitment to protecting Australia's unique forest habitats, in this, the first stage of a global conservation legacy for future generations.

APPROACH

Despite the unique natural and cultural values of the *Eucalyptus* genus of trees, there are currently no conservation reserves in Australia, or its northern neighbours, which are dedicated to protecting the multiple species, or sequences of species, that make up the forest ecosystems of this special member of the *Myrtaceae* family. At the same time, it is essential that those tree-dwelling mammals that inhabit these forests are also conserved. Protecting the manifold expressions of eucalypt forests as they are distributed across their natural range will also look after those plants and animals, of which the Koala is but one, albeit significant, example. Hence this proposal for a Greater Koala Park, as Stage One of the Australian component of a larger, international, *Eucalyptus* Sequence World Heritage Area.

METHOD

The primary method of presenting the information collected is in a series of maps derived from spatial analysis, demonstrating the conservation and habitat values of the proposal area. The datasets consist of the Commonwealth and State Governments' high conservation value (HCV) forest data generated as part of the comprehensive regional assessment undertaken during the New South Wales Regional Forest Agreement (RFA) process, intersected with various State-level data, including forest disturbance, plantation areas and Koala species sightings. The Koala species data have been extracted from the complete inventory of State records as of April 2024 and cover the entire proposal area, using a six-point scale to determine likelihood of presence (low – very high), generated using the NSW Koala Likelihood Map v2.0 (August 2019). Data collected for the flora, fauna and conservation status tables were derived from an aggregation of State, Federal and publicly available sources. The limitations and inherent biases in these data are acknowledged; any associated commentary should be understood as preliminary in nature, and requiring a more comprehensive assessment as part of a formal nomination.

Protecting the remaining Koala forest habitat on the east coast is an absolute priority for Koala conservation nationally

Eucalypt forest, Tuckers Nob

CONSERVATION VALUES

FORESTS

The forests contained within the proposal occupy three climatic zones (subtropical, warm temperate and temperate) and are largely determined by altitude and distance from the coast, as well as aspect, terrain, soil, and availability of water. Each of these factors influence the types of forest (native eucalypt forest or rainforest) as well as individual tree species, and can vary considerably as a consequence of micro-climatic conditions. The coastal eucalypt forests contain extensive areas of coastal blackbutt (*E. pilularis*) but a range of other eucalypts can be present, including Blue gum (*E. saligna*), Flooded Gum (*E. grandis*) and Tallowwood (*E. microcorys*), which are also common plantation species, as well as Small-fruited Grey Gum (*E. propinqua*), Coastal Ironbark (*E. siderophloia*), Swamp-, Red- and White Mahogany (*E. robusta, E. resinifera, E. acmenoides*). Further inland and at higher altitudes other species are to be found, including New England Blackbutt (*E. campanulata*), Messmate (*E. obliqua*), Forest Red Gum (*E. tereticornis*) Mountain Ribbon Gum (*E. nobilis*), as well as the occasional New England- and Narrow-leaved Peppermint (*E. nova-anglica, E. radiata*), and Snow Gum (*E. pauciflora*). Some species are endemic to the region, such as the Dorrigo White Gum (*E. dorrigoensis*). The other two genera commonly ascribed to the eucalypt family, Corymbia (bloodwood) and Angophora (apple) are also present. Various other tree genera, which are found across a range of forest communities, include the Forest Oak (*Allocasuarina torulosa*), River Oak (*Casuarina cunninghamiana*), wattle (*Acacia* spp.), Brush Box (*Lophostemon confertus*) and Turpentine (*Syncarpia glomulifera*).

There are also extensive areas of rainforest throughout the proposal, some of which are already recognised within the formal reserve system as part of the Gondwana Rainforests of Australia World Heritage Area. These include the Antarctic Beech (*Nothofagus moorei*) in the more temperate zones, many of which were heavily impacted by the Black Summer fires, and whose recovery is both slow and uncertain. The unburnt beech forests in the higher-elevation State Forests of the hinterland would consequently be a welcome addition to the conservation estate. Of further significance are the areas of lowland, sub-tropical rainforest, recognised by the Commonwealth Government as critically endangered, and which generally occur at elevations of less than 300 metres, although soil and aspect can result in patches at higher altitudes. The proposal area constitutes the largest remaining stronghold for this forest type; other locations exist, but they are far more fragmented. The number of species which make up the classification are numerous, but some of the more well-known trees include the Hoop Pine (*Araucaria cunninghamii*) as well those distinguished by their buttressed trunks, including the Yellow Carabeen (*Sloanea woollsii*) and the Giant Stinging Tree (*Dendrocnide excelsa*) and the once extensive but now less common Red Cedar (*Toona ciliata*). Figs (*Ficus spp.*) and palms, notably the Bangalow Palm (*Archontophoenix cunninghamiana*) and Cabbage Tree Palm (*Livistona australis*) give the rainforests of the region their distinct appearance, and are found across the warmer climes of south eastern Australia. Other sub-canopy and understorey species are more confined, such as the Rusty Plum (*Niemeyera whitei*), listed as vulnerable in New South Wales, and the Milky Silkpod (*Parsonsia dorrigoensis*) which, with the exception of one small population slightly further south, is found almost exclusively within the proposal area. One final tree of note is the Ringwood or Aniseed Myrtle (*Syzygium anisatum*) found only in the Bellinger, Nambucca and Kalang River valleys, which is now under threat from the introduced myrtle rust (*Puccinia psidii*) and has been submitted to the New South Wales Threatened Species Scientific Committee to be considered for listing as endangered.

Multiple groves and patches of old-growth forest and rainforest exist in the area currently zoned State Forest within the proposal, including the areas zoned plantation. Although expected to be excluded from removal during extractive timber operations, such habitat can be unmapped, overlooked, or misclassified while machinery introduces weeds, brings in disease such as rust, and damages trees. Hollow bearing trees in the public forest estate are crucial for a range of threatened bird species and mammals including the Glossy Black-Cockatoo (*Calyptorhynchus lathami*) listed as vulnerable, and the endangered Greater Glider (*Petauroides volans*). The Grey-headed Flying Fox (*Pteropus poliocephalus*) also vulnerable, is one of the keystone pollinators of the region. Given the age and global significance of many of the species in the proposal area, and the threats they face, conservation is the best form of management available.

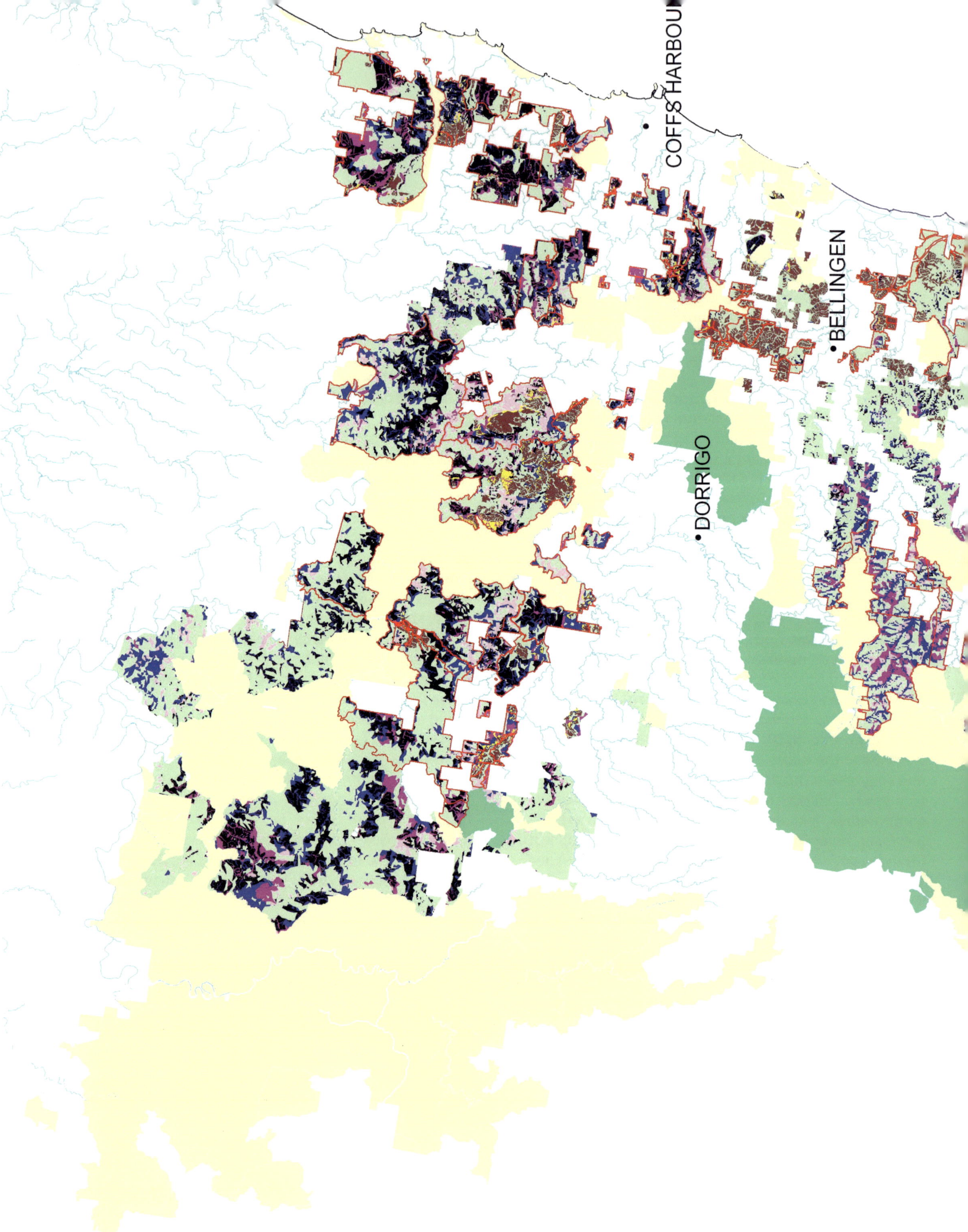
COFFS HARBOU
BELLINGEN
DORRIGO

FIG. 2

Greater Koala Park:

HIGH CONSERVATION VALUES

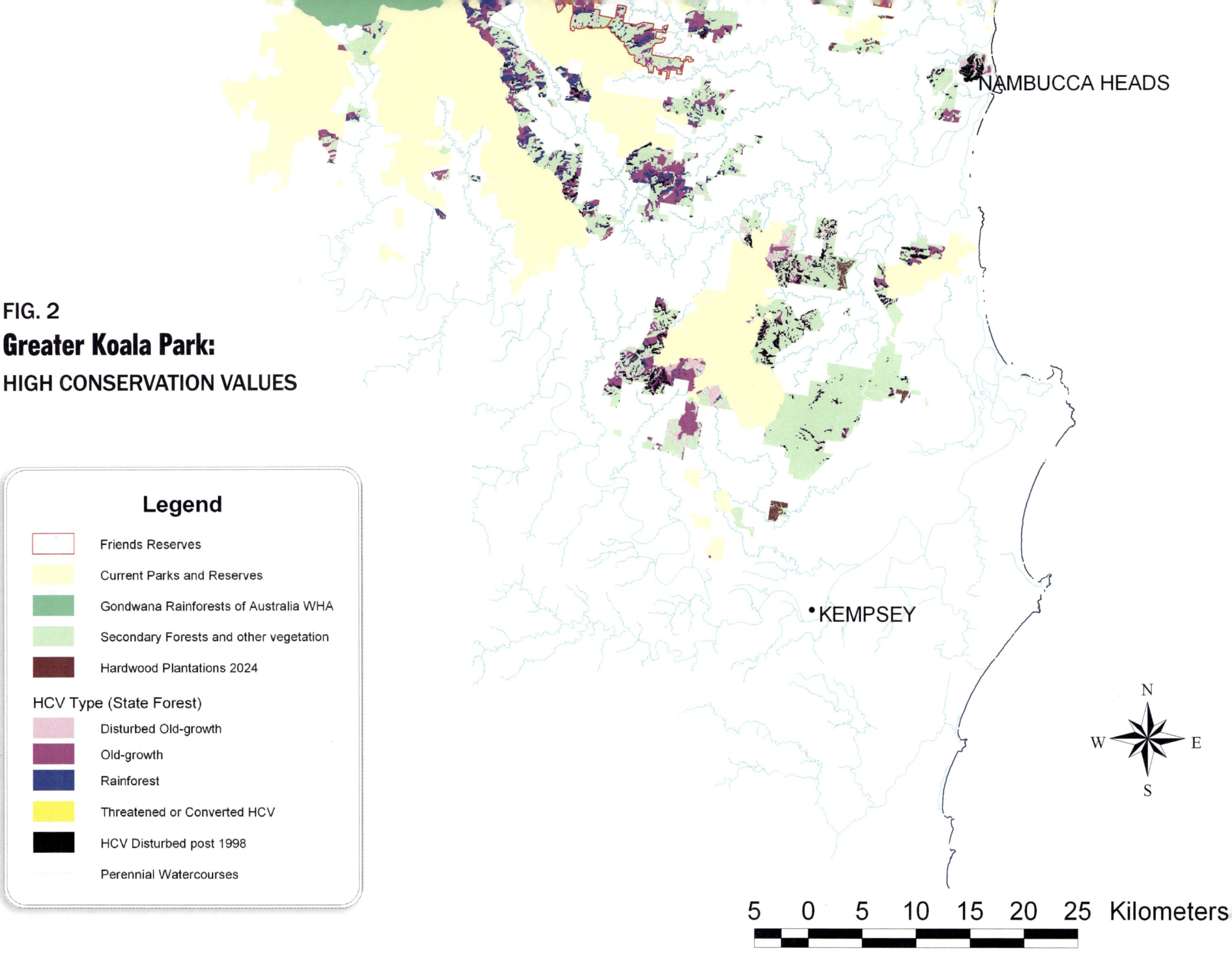

Koala, Tarkeeth

CONSERVATION VALUES

KOALAS

Climate change poses a serious risk to the survival of the Koala, due to increased frequency and intensity of bushfires. Black Summer resulted in the death or injury of as many as sixty thousand individuals. Habitat loss and fragmentation, forestry, urban sprawl, vehicle collisions, dog attacks and stress-induced disease are also taking their toll and without drastic intervention the Koala will be extinct by 2050. Conservation attempts in the early years of the last century were based on capturing and moving wild populations to islands off the coast of southern Australia, a process known as translocation, to be brought back onto the mainland in the future. This method proved successful and there are now translocated Koala populations in South Australia and Victoria that have moved back into available habitat. This history is not the same as the wild populations of southeast Queensland and northern New South Wales that survived over time. It is these "northern" Koalas that inhabit the proposal area.

The significance of the proposed reserve for ensuring the ongoing viability of these wild Koalas cannot be underestimated. Few other locations on the east coast are as sufficiently large or with so much habitat free from human development as this proposal. Koalas cannot survive in small reserves and wildlife corridors and links are not enough. They need habitat of sufficient size and quality with a good range of tree species, growing on a range of soil types, and permanently available sources of water. These are all abundant within the proposal area.

Unfortunately there is a common misunderstanding that Koalas only live in native forests and consequently a belief that plantation forestry can continue within the reserve with minimal impact. This is not the case. Koalas do not discriminate between the different land uses and management practices by which humans classify the landscape. They go to where the food is good and there is room to move as they are largely solitary, except when breeding. Significantly three of the four plantation species (Flooded Gum, Blue Gum, and Tallowwood) are primary feed tree species which is what makes the plantations of the proposal area so attractive to the Koala and so critical for its survival, which is why this proposal does not exclude them from the reserve. The plantations within the proposal area are also located in areas important for the movement of Koalas across the landscape and have been mostly established on moist, productive soils, making them highly desirable habitat. To make matters worse a significant proportion of these areas were not planted forests but areas of native forest regrowth and original forest.

The systematic replacement of these mixed species plantations with the one species which is not especially palatable to Koalas (Coastal Blackbutt) will be highly problematic going forward. Not only will ongoing forestry operations create permanent disruption within the reserve, the Koalas in these areas will be heavily impacted since they will lose both their habitat and the food sources within it, as these plantations are progressively converted to monoculture. A further problem is that these plantation areas are universally located within larger areas of native forests and rainforests, which are also zoned plantation, and have no secure conservation status. A viable park needs viable boundaries and all suitable habitat should be included whatever land use classification it may have. Otherwise, the park will be full of holes and act as a permanent drain on the very population of Koalas it has been designed to protect.

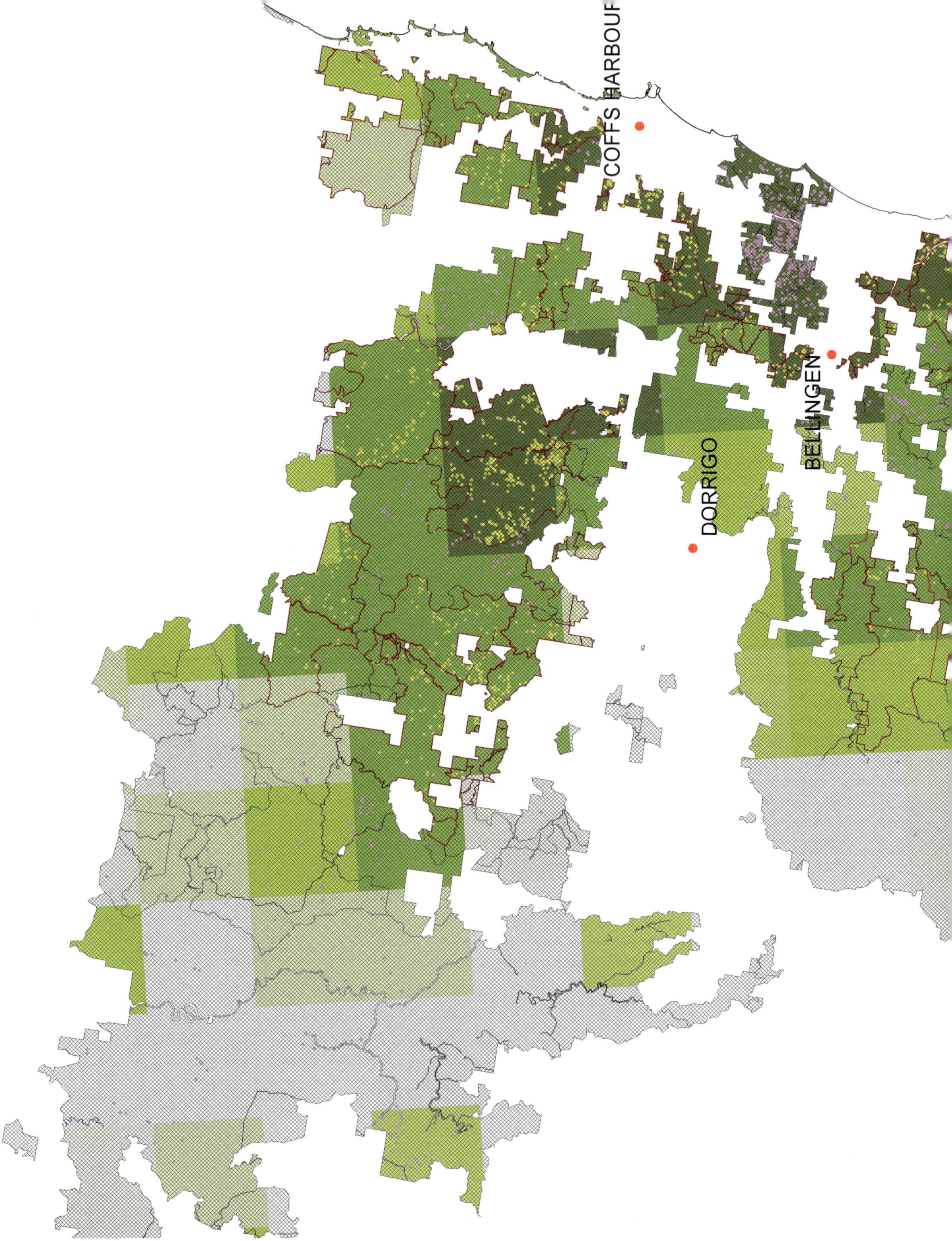
COFFS HARBOU
DORRIGO
BELLINGEN

FIG. 3

Greater Koala Park:

KOALA SPECIES SIGHTINGS

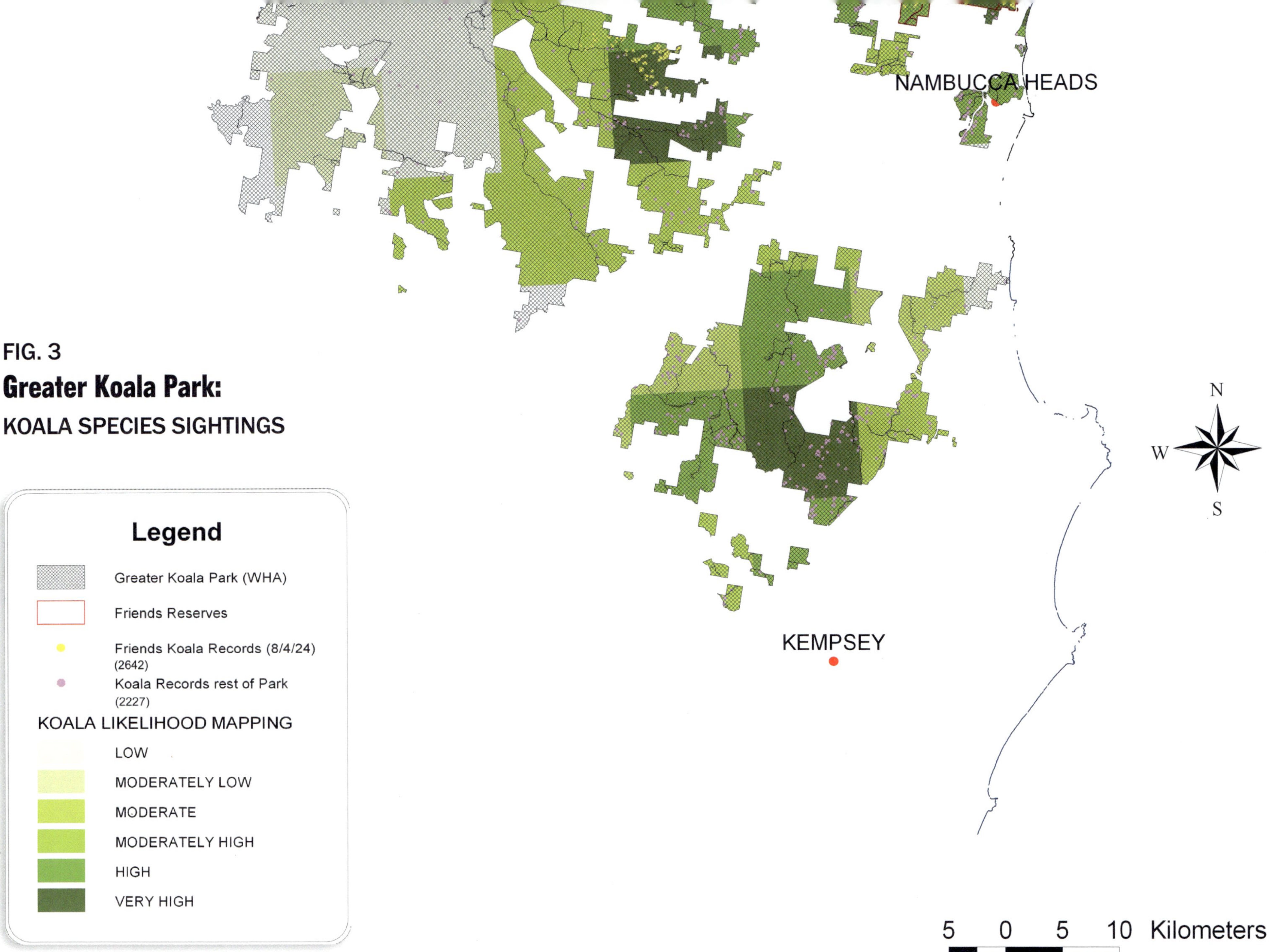

Upper Kalang River

SUMMARY OF WORLD HERITAGE VALUES

Table 1. Criteria for assessment and indicative justification for listing.

	Selection criteria	Value(s)	Eligibility	Indicative justification
(i)	to represent a masterpiece of human creative genius;	Cultural	No	N/A
(ii)	to exhibit an important interchange of human values, over a span of time or within a cultural area of the world, on developments in architecture or technology, monumental arts, town planning, or landscape design;	Cultural and Natural	Yes	Cultural landscape managed by First Nations over millennia
(iii)	to bear a unique or at least exceptional testimony to a cultural tradition or to a civilization that is living or that has disappeared;	Cultural	Yes	Locality of First Nations creation stories and living culture relating to country
(iv)	to be an outstanding example of a type of building, architectural or technological ensemble, or landscape that illustrates (a) significant stage(s) in human history;	Cultural and Natural	Yes	See (ii), (iii) above and (v), (vi) below
(v)	to be an outstanding example of a traditional human settlement, land use, or sea use that is representative of a culture (or cultures) or human interaction with the environment, especially when it has become vulnerable under the impact of irreversible change;	Cultural and Natural	Yes	Locality of First Nations stories and culture relating to changes in country
(vi)	to be directly or tangibly associated with events or living traditions, with ideas, or with beliefs, with artistic and literary works of outstanding universal significance. (The Committee considers that this criterion should preferably be used in conjunction with other criteria);	Cultural	Yes	See (ii), (iii), (iv), (v) above, notably creation stories
(vii)	to contain superlative natural phenomena or areas of exceptional natural beauty and aesthetic importance;	Natural	Yes	Several areas already listed with these values
(viii)	to be outstanding examples representing major stages of earth's history, including the record of life, significant ongoing geological processes in the development of landforms, or significant geomorphic or physiographic features;	Natural	Yes	Several areas already listed due to Gondwana association
(ix)	to be outstanding examples representing significant ongoing ecological and biological processes in the evolution and development of terrestrial, fresh water, coastal, and marine ecosystems and communities of plants and animals;	Natural	Yes	Locality for a diverse array of plants and animals, notably eucalypt species
(x)	to contain the most important and significant natural habitats for in-situ conservation of biological diversity, including those containing threatened species of outstanding universal value from the point of view of science or conservation	Natural	Yes	Known stronghold for endangered species, notably the Koala

(Source: Cadman et al, 2024 - see Further Reading)

FINDINGS

Council Koala sign, Gleniffer

There are several key findings arising from the research undertaken.

The first is the extent to which the HCV forests identified during the Commonwealth-State RFA process of the 1990s remain within the proposal area. Approximately one quarter of the State Forest is still HCV, with considerable areas of rainforest, old-growth forest and disturbed old-growth forest remaining to this day. Unfortunately, more than fifteen per cent of the HCV forest that existed at that time has been subjected to forestry operations.

A second and related finding concerns the levels of native vegetation and HCV forest within the areas currently zoned plantation. At least a quarter continues to be dominated by native vegetation, while the research identified a further one tenth of the area as HCV, which has either been converted to plantation or is under threat of conversion to plantation.

Koalas are to be found across the proposal area with heavy concentrations in the coastal lowland forests. There are fewer in the national parks than state forests and records become sparser further west and in the higher elevations, with the notable exception of the Dorrigo plateau and hinterlands. Whether this is an artefact of the data or an indication of a broader trend is unclear but it certainly emphasises the relevance of including the state forest in the conservation estate.

It is also important to consider the cultural as well as the natural values against the criteria for World Heritage. The Gumbaynggirr peoples have a traditional and ongoing relationship with the plants and animals within the proposal area. Ensuring the continuance of this cultural association through opportunities to care for Country with a formal role in land management will be central to a successful nomination.

While all forests are important the significance of the Friends reserves is clear. The majority of Koalas and high conservation value forests are to be found within them and the high to very high likelihood of Koalas in these areas further underscores their importance. The fact that most of the areas zoned plantation are also found within them is no coincidence. It is an inconvenient, albeit unavoidable truth, that Koalas do not discriminate between natural and planted forests. Rather, they go where food, water and habitat is available. The diversity of forest types and ecosystems makes them critical to the conservation of the flora and fauna of the Mid North Coast.

RECOMMENDATIONS

Although the Regional Forest Agreement data are over a quarter of a century old, they remain the only source of forest-related information jointly validated by both Federal and State governments. Many refugia and micro-habitats were left undetected and unprotected, a situation which continues to the present. Many state agencies still rely on these data and continue to overlook areas left out of the original assessment of conservation values.

It is therefore recommended that the Federal and State Governments commit to the generation of new, jointly-validated spatial information in the assessment of the natural and cultural values of the proposed World Heritage area.

In more recent years technologies such as orthorectification and georeferencing have made this historical imagery usable for verification in the field. The creation of new imagery is helpful in this regard but if this is not checked against historical photographic records, and on the ground, such work will also be incomplete.

It is further recommended that existing historical aerial photography be re-visited and re-interpreted to ensure contemporary information is triangulated with past land-use, at both a desktop level, and in the field.

A different set of problems confront the use of Koala records. Many of these records were generated prior to forestry operations and do not reflect subsequent ecological conditions. Areas zoned plantation have been under-sampled, and other localities over-sampled. Many Koala sightings are from, or close to, roads and remote areas or rugged terrain have been overlooked. The Black Summer bushfires have also impacted Koala populations, and these records will be out of date. It is noted that research is currently being undertaken using aerial drones but areas zoned plantation been excluded, creating a bias in the records, thereby making them unreliable for any future assessment for World Heritage.

No areas within the footprint of the proposal should be excluded from assessment for Koalas. Comprehensive research is required to more accurately determine Koala numbers within, and surrounding, the proposal area prior to any nomination for World Heritage listing.

The exclusion of assessment of forest areas zoned plantation under current State Government processes for reservation as a national park are noted. It is important to recognise that many of these areas were never planted,

Tree Goanna, Tarkeeth

contain natural forest and native vegetation, or natural regeneration has outcompeted the planted forest. In some instances these areas are of higher conservation value and better representative examples of eucalypt sequences, than other, depauperate, areas zoned native forest.

No areas within the footprint of the proposal should be excluded from assessment for conservation values. Further research is required to more accurately determine these areas within the proposal, prior to any nomination for World Heritage listing.

Spatial analysis shows ongoing disturbance of conservation values as a result of forestry operations, concommitant deforestation and forest degradation, and fire. The Friends estimate restoration costs at between $5,000 - $20,000 per hectare, and an upper limit of 240 person hours per hectare.

Assessment of the areas to be restored should be based on ecological condition, depending on intensity of previous logging, weed infestation, severity of burning, and levels of erosion. Trained conservation volunteers could offset some costs.

The meaningful participation of all relevant stakeholders in deliberations around environmental, social and economic decision-making is a hallmark of sustainability. It fosters good community relations and engenders ownership of any decisions made.

Future governance arrangements for World Heritage assessment require the full and effective participation of interested and affected parties as well as all other relevant stakeholders.

TABLE DATA

Upper Kalang River

Table 2. Parkwide data.

PARKWIDE	
Total Recorded Koala Sightings	4,869
Total area of reserve proposal (ha)	491,818
National Park and Nature Reserves, including World Heritage	290,017
World Heritage	42,350
Other reserves	1,304
State Forest	200,497
State Forests	Bagawa, Boambee, Boundary Creek, Buckra Bendinni, Chaelundi, Clouds Creek, Collombatti, Conglomerate, Diehappy, Ellis, Gladstone, Gundar, Hyland, Ingalba, Irishman, Kangaroo River, Little Newry, Lower Bucca, Marara, Marengo, Mistake, Moonpar, Nambucca, Nana Creek, Never Never, Newry, Nulla-Five Day, Oakes, Old Station, Orara East, Orara West, Pee Dee, Pine Creek, Roses Creek, Scotchman, Sheas Nob, Skillion Flat, Tamban, Tarkeeth, Thumb Creek, Tuckers Nob, Viewmont, Way Way, Wedding Bells, Wild Cattle Creek
Area zoned Plantation (ha)	15,809
Area zoned hardwood plantation	11,066
Softwood plantation	888
High conservation values (ha)	51,016
Rainforest	22,282
Oldgrowth	16,537
Disturbed oldgrowth	12,197
Other native vegetation in area zoned plantation	3,281
Total disturbed HCV since 1998 (ha)	30,738
Threatened or converted HCV in area zoned plantation (ha)	1,415

Table 3. Friends reserves data.

FRIENDS	
Name of Group:	All Friends
Total Recorded Koala Sightings	2,642
Total area of reserves in proposal (ha)	102,665
State Forest	102,665
State Forests	Boambee, Buckra Bendinni, Clouds Creek, Conglomerate, Diehappy, Ellis, Gundar, Irishman, Kangaroo River, Little Newry, Lower Bucca, Marengo, Moonpar, Nana Creek, Newry, Oakes, , Orara East, Orara West, Roses Creek, Scotchman, Sheas Nob, Tarkeeth, Tuckers Nob, , Wedding Bells, Wild Cattle Creek
Area zoned Plantation (ha)	11,888
Area zoned hardwood plantation	8,564
Softwood	888
High conservation values (ha)	28,982
Rainforest	12,914
Oldgrowth	8,311
Disturbed old-growth	7,757
Total disturbed HCV since 1998 (ha)	19,218

Table 4. Other proposal data.

OTHER	
Total recorded Koala sightings (rest of proposal)	2,227
State Forest (ha)	99,250
State Forests	Bagawa, Boundary Creek, Chaelundi, Collombatti, Gladstone, Hyland, Ingalba, Marara, Mistake, Nambucca, Never Never, Nulla-Five Day, Old Station, Pee Dee, Pine Creek, Skillion Flat, Tamban, Thumb Creek, Viewmont, Way Way
Area zoned Plantation (ha)	3,921
Area zoned hardwood plantation	2,502
High conservation values (ha)	22,034
Rainforest	9,368
Old-growth	8,226
Disturbed old-growth	4,440
Total disturbed HCV since 1998 (ha)	11,520

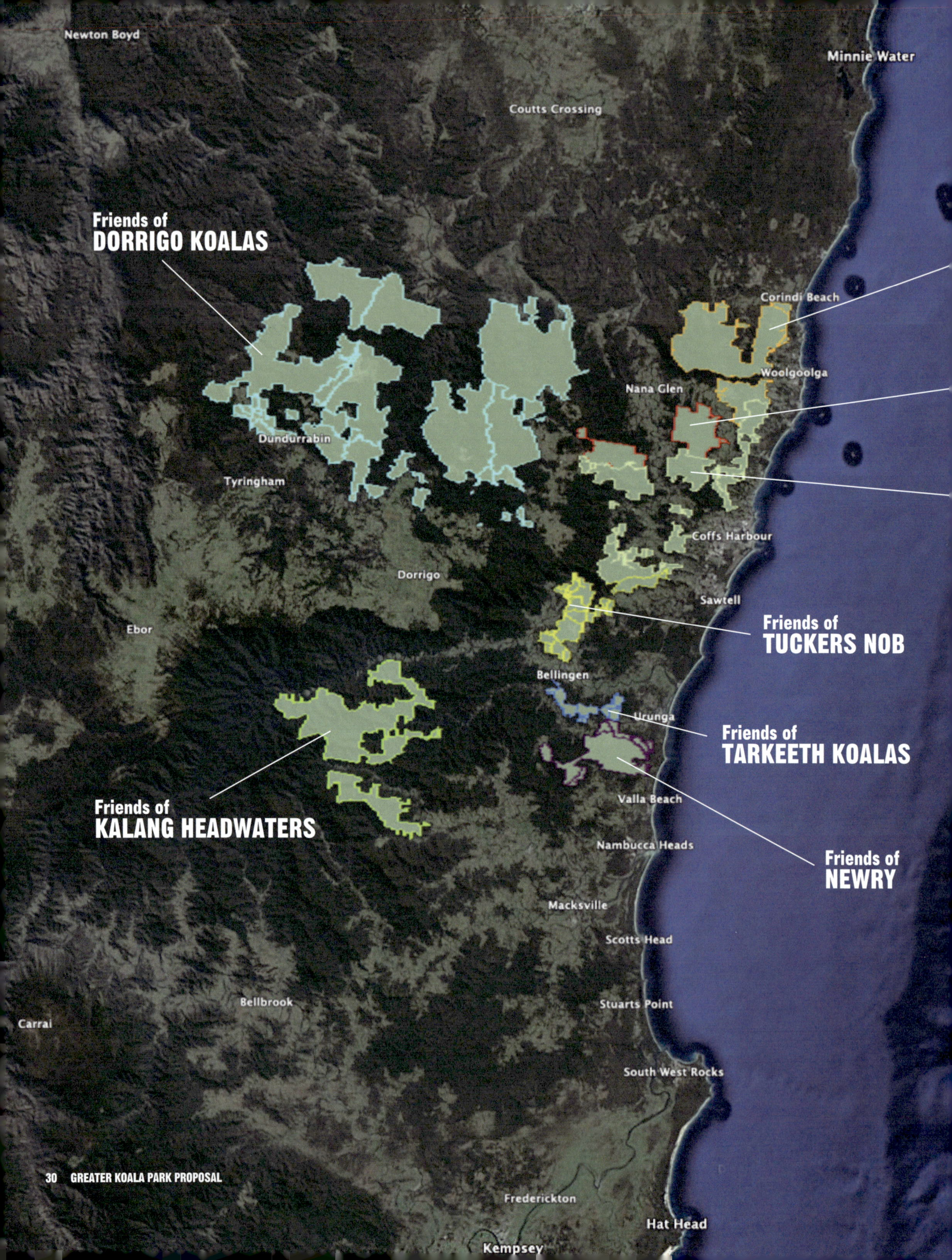
Newton Boyd
Minnie Water
Coutts Crossing
Friends of
DORRIGO KOALAS
Corindi Beach
Woolgoolga
Nana Glen
Dundurrabin
Tyringham
Coffs Harbour
Dorrigo
Sawtell
Ebor
Friends of
TUCKERS NOB
Bellingen
Urunga
Friends of
TARKEETH KOALAS
Friends of
KALANG HEADWATERS
Valla Beach
Nambucca Heads
Friends of
NEWRY
Macksville
Scotts Head
Bellbrook
Stuarts Point
Carrai
South West Rocks
Frederickton
Hat Head
Kempsey

FRIENDS

RESERVE PROPOSALS

We support the establishment of the Greater Koala Park World Heritage Area that connects vital Koala habitat across the NSW Mid North Coast. This initiative aims to safeguard Koala populations and enhance ecological resilience

FRIENDS OF
COFFS COAST KOALAS

PHILOSOPHY

The Friends of Coffs Coast Koalas is an alliance of local community members who live on the Coffs Coast, care passionately about Koalas and want to protect them from extinction.

Our philosopy is to preserve the unique forests of Coffs Harbour as refuges for the diversity of local native species to survive and thrive. Our focus is on the public forests of Nana Creek, Lower Bucca and Gundar.

We are acutely aware of the destructive impacts of increased industrial logging, over-development and climate change on public native and plantation state forests. We are concerned for the loss of forest habitat that is directly impacting the Koalas chance for survival beyond 2050.

MISSION

- To create refuges of connected and integrated forest to foster biodiversity, save the Koala and other endangered and threatened species.
- To protect public forests to mitigate the impacts of human expansion, industry and climate change and in doing so promote water security, carbon sequestration and species survival.
- We urgently call for an end to industrial logging in public forests.
- To regenerate depleted landscapes as areas of refuge for the endangered Koala.
- To support local and international initatives to provide greater protection for these unique and world renowned forests.
- To utilise current science and methodology.
- To promote community awareness of the Koalas plight and encourage greater protection for the unique eucalypt forests that support a flourishing wild Koala population.

GOALS

- To promote awareness through public campaigns that highlight the destructive practices of past and current industrial logging and over-development.
- To engage in community meetings, social media and local outreach; our goal is to spread awareness of the issues impacting the survival of the Koala in our region.
- To remain apolitical; putting the interests of the forests and species survival over party politics
- To create a welcoming and inclusive environment to encourage people to join our concerns for the survival of the endangered Koala.
- To respect the traditional lands of the Gumbaynggirr Nation and support initiatives for greater sovereignty and cultural practices on Country.
- To share information with Friends groups and support community initiatives to protect landscape connectivity.
- To support the establishment of the Greater Koala Park World Heritage Area (GKPWHA) that connects vital Koala habitat across the NSW Mid North Coast. This initiative aims to safeguard Koala populations and enhance ecological resilience.

Nana Creek

FRIENDS OF COFFS COAST KOALAS
RESERVE DETAILS

Table 5a. FoCCK - Reserve data.

Name of Group:	**FRIENDS OF COFFS COAST KOALAS**
Total Recorded Koala Sightings	**220**
Total area of reserve (ha)	**4,722**
State Forest	**4,722**
State Forests	**GUNDAR, LOWER BUCCA, NANA CREEK**
Area zoned Plantation (ha)	**237**
Area zoned hardwood plantation	**196**
High conservation values (ha)	**1,194**
Rainforest	**453**
Old-growth	**624**
Disturbed old-growth	**117**
Total disturbed HCV since 1998 (ha)	**2,031**

Indicative assessment area of restoration (ha): 2,500

In their current states of regrowth and recovery, and with no further logging, the resilience of these forests should see healthy natural regeneration with very limited assisted regeneration, with the exception of compartments 003 and 004 in Lower Bucca, and Gundar, where some areas of HCV forest and Koala habitat have been heavilly distrubed in recent years.

Type of restoration:

Depending on the degree of weed infestation or other degradation e.g. machinery, may be required for initial clearing, thinning, ground preparation or remediation work. A minimum of 3 years annual follow-up work is required, ideally 10 years. Savings can be made with volunteer input and by sourcing donations of plants, materials, expertise and with grant funding.

FRIENDS OF COFFS COAST KOALAS

PROPOSED RESERVE

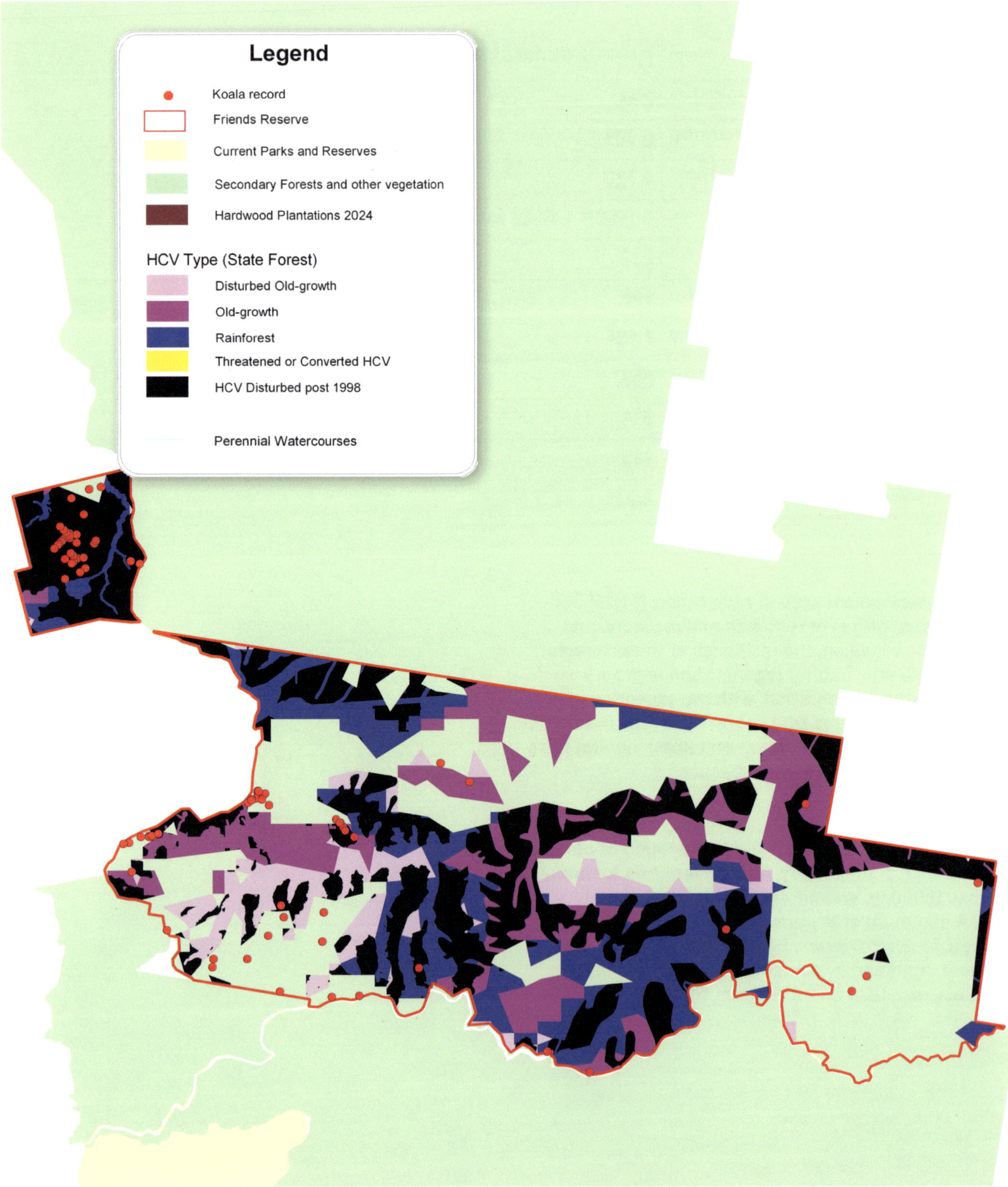

Fig. 4 FoCCK - Proposed reserve

FRIENDS OF COFFS COAST KOALAS
RESERVE DETAILS

FOREST HABITATS

Eucalypt and rainforest plant communities dominate the landscape. The sheer diversity of fire dependent eucalypt species coupled with fire sensitive rainforest understory is in itself an internationally unique ecosystem. The presence of high conservation old-growth forests, dry and wet sclerophyll forests, sub-tropical lowland rainforest, subtropical coastal floodplain forest (Endangered Ecological Communities or EEC) further qualifies these forests as areas of international significance.

A range of habitat types are represented within a relatively compact area in Lower Bucca, where coastal swamps on sand quickly change to steep foothills rising toward the Mount Coramba peak at 590mASL.

Lowland, hinterland and escarpment wet sclerophyll forests are represented, along with dry rainforest, lowland subtropical rainforest and freshwater wetlands.

Comparatively small areas of dry sclerophyll forest are present (less than 10%) on low-lying sands, and some of the species representations here are indicative of swamp sclerophyll forest. Patches of old-growth provide crucial habitat for many species. The near total absence of dry sclerophyll plant communities in all three forests suggests these forests are excellent fire and drought refuges for threatened, and other, species.

Primary and secondary Koala habitat is present throughout the landscape. Areas of secondary Koala habitat are important to juvenile and dispersing Koalas, provide seasonal and drought foraging habitat and act as fire refuges.

The eucalypt forests of the North Coast of NSW have been recognised as some of the most diverse and species rich communities in the world. These forests include high levels of both endemic eucalypt diversity and overall eucalypt diversity.

Table 5b. Koala food trees present and regional use rate.

Tallowwood *Eucalyptus microcorys*	High
Small-fruited Grey Gum *Eucalyptus propinqua*	High
Swamp Mahogany *Eucalyptus robusta*	High
Forest Red Gum *Eucalyptus tereticornis*	High
Red Mahogany *Eucalyptus resinifera*	High
Sydney Blue Gum *Eucalyptus saligna*	Significant
Flooded Gum *Eucalyptus grandis*	Significant
Turpentine *Syncarpia glomulifera*	Significant
Forest Oak *Allocasuarina torulosa*	Significant
Brush Box *Lophostemon confertus*	Occasional
Swamp Box *Lophostemon suaveolens*	Occasional
Pink Bloodwood *Corymbia intermedia*	Occasional
Red Bloodwood *Corymbia gummifera*	Occasional
Broadleaf Paperbark *Melaleuca quinquinervia*	Occasional

THREATENED SPECIES

See Table 5c. FoCCK - Threatened species

REASONS FOR INCLUSION

LANDSCAPE

The internationally significant coastal plains and foothills forests of Coffs Harbour are unique, providing regional Koala population boundaries formed by sandstone-based habitats in the north and lowland subtropical rainforest habitats to the east.

The coastal foothills are dominated by wet and dry sclerophyll communities with diverse towering Eucalypt forests. These eucalypt forests are historically unique in that no other genus of trees in the world dominate the vegetation landscape over such vast terrains. FoCCK recognises the importance of eucalypts as part of the global natural heritage.

There are equally significant threatened ecological communities present within these forests including the remarkable Endangered Ecological Communities (EEC) of 'Lowland Rainforest in the NSW North Coast and Sydney Basin Bioregions'.

Nana Creek waterfall

KOALAS

Connectivity is key; Nana Creek, Lower Bucca and Gundar forest areas contain some of the Coffs Harbour - Guy Fawkes Koala meta-population. These forests are adjacent to the magnificent Bagawa forest, forming a vital regional Koala corridor linking remnant and fragmented habitat. The expanse of the Lower Bucca forests connect vital Koala habitat corridors from Ulidarra NP and the meta Koala hub populations of Orara East forest.

The Greater Koala Park will safeguard these unique forests creating a Koala corridor superhighway and act as an interchange for dispersing Koalas from two crucial regional populations, the Coffs Harbour and Guy Fawkes meta-populations.

CULTURAL HERITAGE

FoCCK acknowledges the significant cultural heritage existing within the reserve's boundaries, and respects the ongoing cultural connection to the flora and fauna across the proposal area of the Gumbaynggir Nation. We support Indigenous-led ecosystem management on Country.

TOURISM AND RECREATION

The exquisite natural beauty of these unique forests with their ancient Gumbaynggirr cultural heritage, old-growth forests, cascading waterfalls, towering eucalypt forests, crystal clear creeks, spectacular vistas, and iconic Australian animals provide a plethora of tourist opportunities. Today, locals and tourists flock to these natural wonderlands for bushwalking, bird watching, mountain biking, forest bathing, and camping as well as providing the basis for ecological scientific research.

The added value of the Greater Koala Park will enhance Coffs Harbour's status as NSWs first eco-tourist destination and bring the international community to its doorstep. It is our hope that the plight of the endangered Koala will gain significant recognition through the added exposure of increased tourism to the area.

WORLD HERITAGE VALUES

Koalas are an internationally iconic, endangered and globally significant species. The NSW Mid North Coast is an important stronghold for Koalas. The Koala is one of the world's most loved and adored species worthy of every possible protection.

The uniqueness of Australia's east coast eucalypt species is evident in the astounding diversity of species and the title as one of the world's fastest growing trees. Eucalypts are the climate crisis saviours and are the worlds vegetative heroes for carbon sequestration. Eucalypts are unique for their ability to thrive in almost all of the diverse geographical landscapes in Australia. From the small shrub species in the north to the towering giants of the south they are truly the world's most adaptive tree species.

Rainforest threatened ecosystems (EEC, TEC) exist within the understory of the eucalypt forests on the Mid North Coast, a symbiotic relationship found nowhere else on earth.

FRIENDS OF COFFS COAST KOALAS
RESERVE DETAILS

Nana Creek

CURRENT ISSUES

Anthropocene barriers - Intensified logging and urban development are increasingly impeding Koala movement and degrading and destroying habitat, affecting population stability, gene flow and food accessibility. Re-establishment of landscape-level linkages are urgently needed for the survival of this endangered species.

Water security / threatened species - Significant areas of High Conservation old-growth forest and patches of Subtropical Lowland Rainforest are interspersed throughout these forests. Typically for North Coast wet forests in the Coffs Harbour region there are numerous streams that, from these forests, feed the mighty Orara River, home to the endangered Eastern Freshwater Cod and part of the Clarence River Catchment which has the only stable population of this endangered fish. Protection of these forests will aid in threatened species survival, ensure future water security and provide refuge and sanctum in climate extremes.

Rainforest

FUTURE VISION

The Friends Of Coffs Coast Koalas support the urgent establishment of the Greater Koala Park World Heritage Area. This globally significant initiative represents an important step towards securing the future of Koalas, and the broader ecosystems upon which this endangered species depends to survive and thrive. It aligns with an urgent commitment towards biodiversity preservation, climate change mitigation, and sustainable tourism.

We urge stakeholders to support this proposal, recognizing the ecological, cultural, and economic value it will bring to the region. Together, we can ensure a brighter future for Koalas living wild and healthy, protected in the towering eucalypt forests of NSW Mid North Coast for all the world to treasure.

This globally significant initiative represents an important step towards securing the future of Koalas and the broader ecosystems upon which this endangered species depends

High conservation value forest, FoCCK reserve

THREATENED SPECIES

Table 5c. FoCCK - Threatened species

COMMON NAME	SCIENTIFIC NAME	COMMONWEALTH ENVIRONMENT PROTECTION & BIODIVERSITY CONSERVATION ACT STATUS	NSW BIODIVERSITY CONSERVATION ACT STATUS
MAMMALS			
Koala	*Phascolarctos cinereus*	Commonwealth Endangered	NSW Endangered
Long-nosed Potoroo	*Potorous tridactylus*	Comm End	NSW Vulnerable
Brush-tailed Rock Wallaby	*Petrogale penicillata*	Commonwealth Vulnerable	NSW End
Rufous Bettong	*Aepyprymnus rufescens*		NSW Vul
Greater Glider	*Petauroides volans*	Comm End	NSW End
Yellow-bellied Glider	*Petaurus australis*	Comm Vul	NSW Vul
Squirrel Glider	*Petaurus norfolcensis*		NSW Vul
Eastern Long-eared Bat	*Nyctophylus corbeni*	Comm Vul	
Grey-headed Flying Fox	*Pteropus poliocephalus*	Comm Vul	NSW Vul
Large-eared Pied Bat	*Chalinolobus dwyeri*	Comm Vul	NSW Vul
Golden-tipped Bat	*Kerivoula papuensis*		NSW Vul
Greater Broad-nosed Bat	*Scoteanax rueppellii*		NSW Vul
Spotted-tailed Quoll	*Dasyurus maculatus*	Comm End	NSW Vul
Common Blossom Bat	*Syconycteris australis*		NSW Vul
Southern Myotis	*Myotis macropus*		NSW Vul
BIRDS			
Glossy Black-Cockatoo	*Calyptorhynchus lathami*	Comm Vul	NSW Vul
Little Lorikeet	*Glossopsitta pusilla*		NSW Vul
Wompoo Fruit-dove	*Ptllinopus magnificus*		NSW Vul
Powerful Owl	*Ninox strenua*		NSW Vul
Sooty Owl	*Tyto tenebricosa*		NSW Vul
Masked Owl	*Tyto novaehollandiae*		NSW Vul
Barking Owl	*Ninox connivens*		NSW Vul
White-bellied Sea Eagle	*Haliaeetus leucogaster*		NSW Vul
Red Goshawk	*Erythrotriorchis radiatus*	Comm Vul	NSW Critically Endangered
Regent Honeyeater	*Anthochaera phrygia*	Commonwealth Critically Endangered	NSW Crit End
Swift Parrot	*Lathamus discolor*	Comm Crit End	NSW End
Varied Sitella	*Daphoenositta chrysoptera*		NSW Vul
White-throated Needletail	*Hirundapus caudacutus*	Comm Vul	NSW Vul
AMPHIBIANS			
Giant Barred Frog	*Mixophyes iteratus*	Comm Vul	NSW Vul
Stuttering Frog	*Mixophyes balbus*	Comm Vul	NSW End

Sphagnum Frog	*Philoria sphagnicolus*	Comm Vul	NSW Vul
Pouched frog	*Assa darlingtoni*		NSW Vul
REPTILES			
Stephens' Banded Snake	*Hoplocephalus stephensii*		NSW Vul
FISH			
Eastern Freshwater Cod	*Maccullochella ikei*	Comm End	NSW End
FLORA			
Cryptic Forest Twiner	*Tylophora woolsii*	Comm End	NSW End
Dorrigo Daisy Bush	*Olearia flocktoniae*	Comm End	NSW End
Milky Silkpod.	*Parsonsia dorrigoensis*	Comm End	NSW Vul
Moonee Quassia	*Quassia spp Moonee Creek.*	Comm End	NSW End
Native Guava	*Rhodamnia rubescens*	Comm Crit End	NSW Crit End
Orara Boronia	*Boronia umbellata*		NSW Vul
Stink Lily	*Typhonium* sp. aff. *brownii.*		NSW End
Rainforest Cassia	*Senna acclinis*		NSW Crit End
Ravine Orchid	*Sarcochilus fitzgeraldii*	Comm Vul	NSW Vul
Red Boppel Nut	*Hicksbeachia pinnatifolia*	Comm Vul	NSW Vul
Red-flowered King of the Fairies	*Oberonia titania*		NSW Vul
Rusty Plum	*Niemeyera whitei*		NSW Vul
Scrub Turpentlne	*Rhodamnia rubescens*	Comm Crit End	NSW Crit End
Slender Marsdenia	*Marsdenia longiloba*	Comm Vul	NSW End
Slender Screw Fern	*Lindsaea incisa*		NSW End
Square-fruited Ironbark	*Eucalyptus tetrapleura*	Comm Vul	NSW Vul
Yellow-flowered King of the Fairies	*Oberonia complanata*		NSW Vul
THREATENED ECOLOGICAL COMMUNITIES			
Lowland Rainforest of Subtropical Australia		Comm Crit End	
Lowland Rainforest on floodplain in the NSW North Coast Bioregion			Endangered Ecological Community
Lowland Rainforest in the NSW North Coast and Sydney Basin Bioregions			ECC
Subtropical Coastal Floodplain Forest of the New South Wales North Coast Bioregion			ECC

To save these forests with all their rich complexity, climate moderating influences and water preserving capabilities from conversion to depauperate dry forests and monocultures

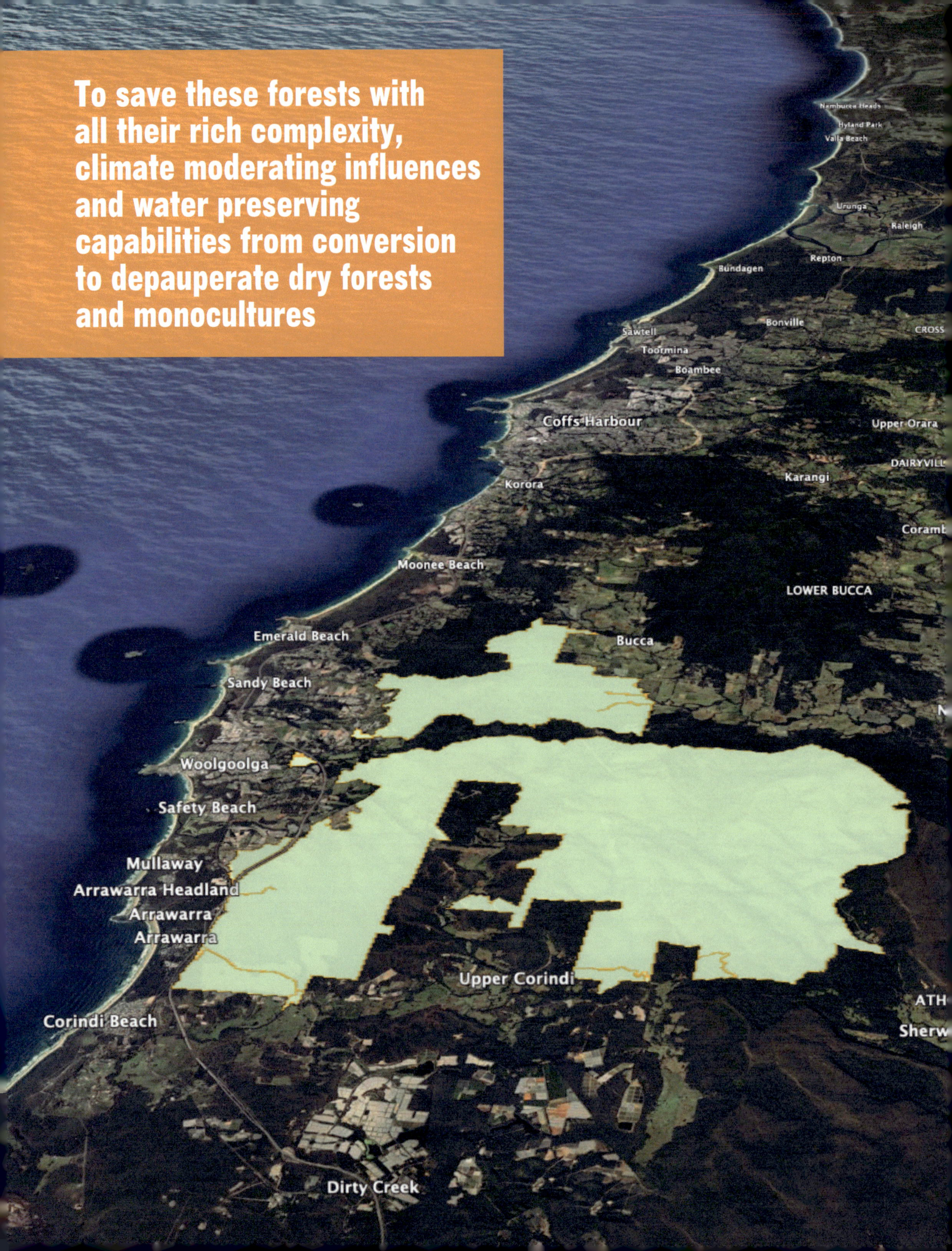

FRIENDS OF CONGLOMERATE

VALUES STATEMENT

PHILOSOPHY

Friends of Conglomerate is a community group of citizen scientists formed in response to witnessing the destruction wreaked upon these extraordinary landscapes by current logging practices including the loss of native forest habitat, wetland, old-growth and rainforest to an ever-expanding 'plantation estate' and unsustainable demand for timber.

We are interested in forest ecology, habitat, species diversity, hydrology, geology and the preservation and protection of forests and their biodiversity and cultural heritage values for the benefit of future generations of all species, with a focus on the forest areas of Conglomerate and Wedding Bells.

MISSION

- To save these forests with all their rich complexity, climate moderating influences and water preserving capabilities from conversion to depauperate dry forests and monocultures
- To save and protect valuable remnants of high conservation value old-growth forest, rainforest, and tall wet sclerophyll forest and the rich diversity of species they support
- To call for an immediate end to logging in public forests
- To return degraded landscapes to vibrant complex forests with healthy, functioning ecosystems and high value habitat
- To promote and perpetuate an increase in the populations of Koala and other threatened species by calling for inclusion in a wider, permanently protected area.

GOALS

- Draw attention to the plight of wildlife and the current loss of species, habitat and forest values caused by clearing and degradation primarily as a result of modern logging practices and exacerbated by climate change
- Include the reserve within the Indigenous Protected Areas system
- Gain knowledge and information needed to promote our cause by sharing information, supporting and networking with forest neighbours, other forest Friends groups and the wider community; by surveying, record keeping and reporting and by organising informative community events to improve and disseminate knowledge, skills and methods
- To remain apolitical while communicating our concerns to relevant Ministers and others in a position to effect positive permanent change.

Sherwood Creek

FRIENDS OF CONGLOMERATE
RESERVE DETAILS

Table 6a. FoC - Reserve data.

Name of Group:	**FRIENDS OF CONGLOMERATE**
Total Recorded Koala Sightings	**24**
Total area of reserve (ha)	**10,672**
State Forest	**10,672**
State Forests	**CONGLOMERATE, WEDDING BELLS**
Area zoned Plantation (ha)	**952**
Area zoned hardwood plantation	**827**
High conservation values (ha)	**2,560**
Rainforest	**854**
Old-growth	**1,302**
Disturbed old-growth	**404**
Total disturbed HCV since 1998 (ha)	**3039**

Indicative assessment area of restoration (ha): 4,050
Plantation - full restoration
CIFOA (NF) - assisted regeneration
Other (old-growth, rainforest, wetland) - assisted regeneration

Type of restoration:
Some parts of the forest have been subjected to intensive forest management regimes, including clear-fall, and partial-clear-fall, as well as escaped renegeration burns and wildfire. Renegetation levels will be pependent on the level of weed infestation and/or site preparation works required, e.g. if machinery is needed for initial clearing of weeds or for other preparatory ground works. A minimum 3 years follow-up maintenance work is necessary, ideally continuing for 10 years. Savings can be made with volunteers contributing labour, materials and expertise, and by sourcing donations and grant funding.

FRIENDS OF CONGLOMERATE
PROPOSED RESERVE

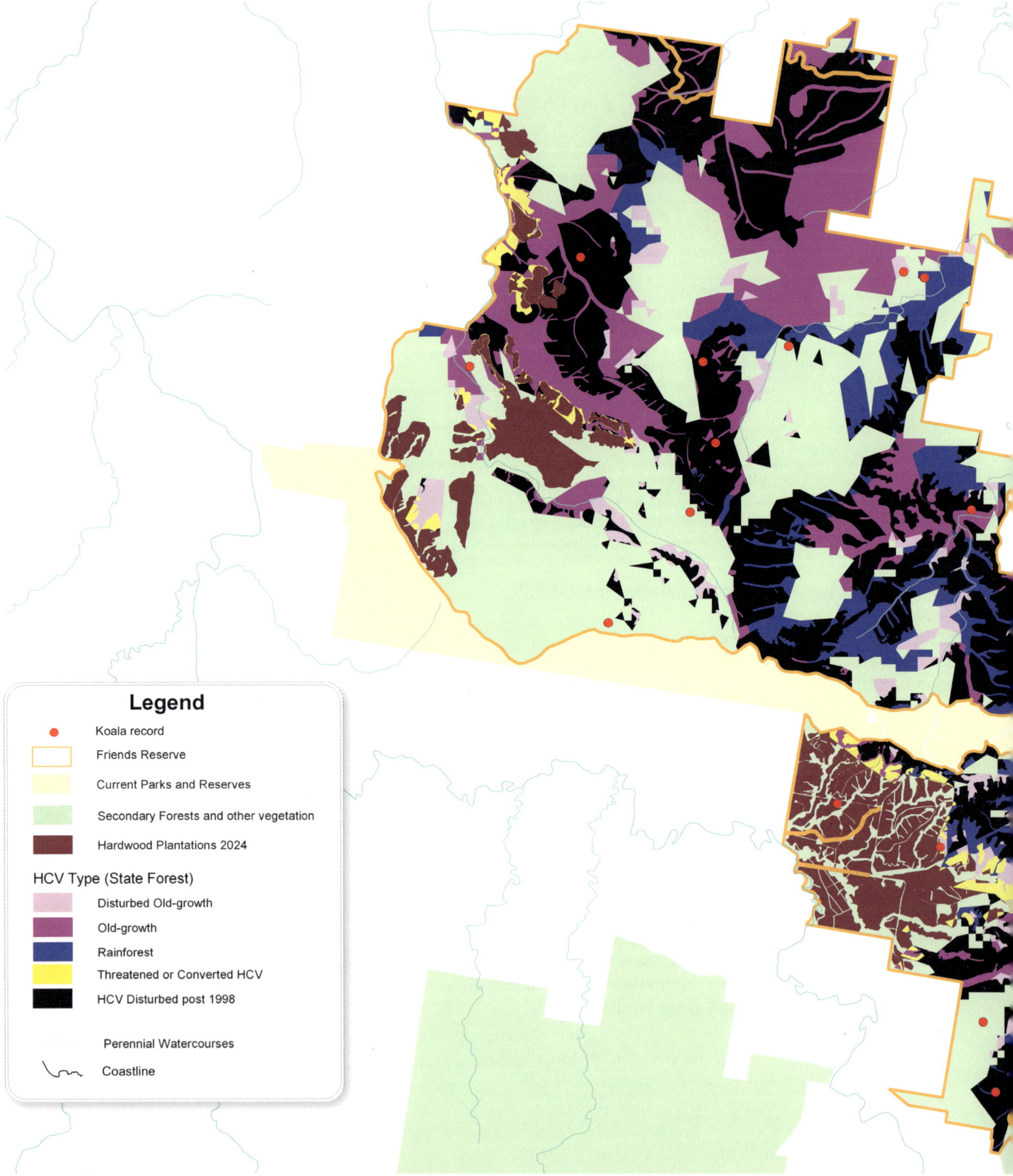

Fig. 5 FoC - Proposed reserve

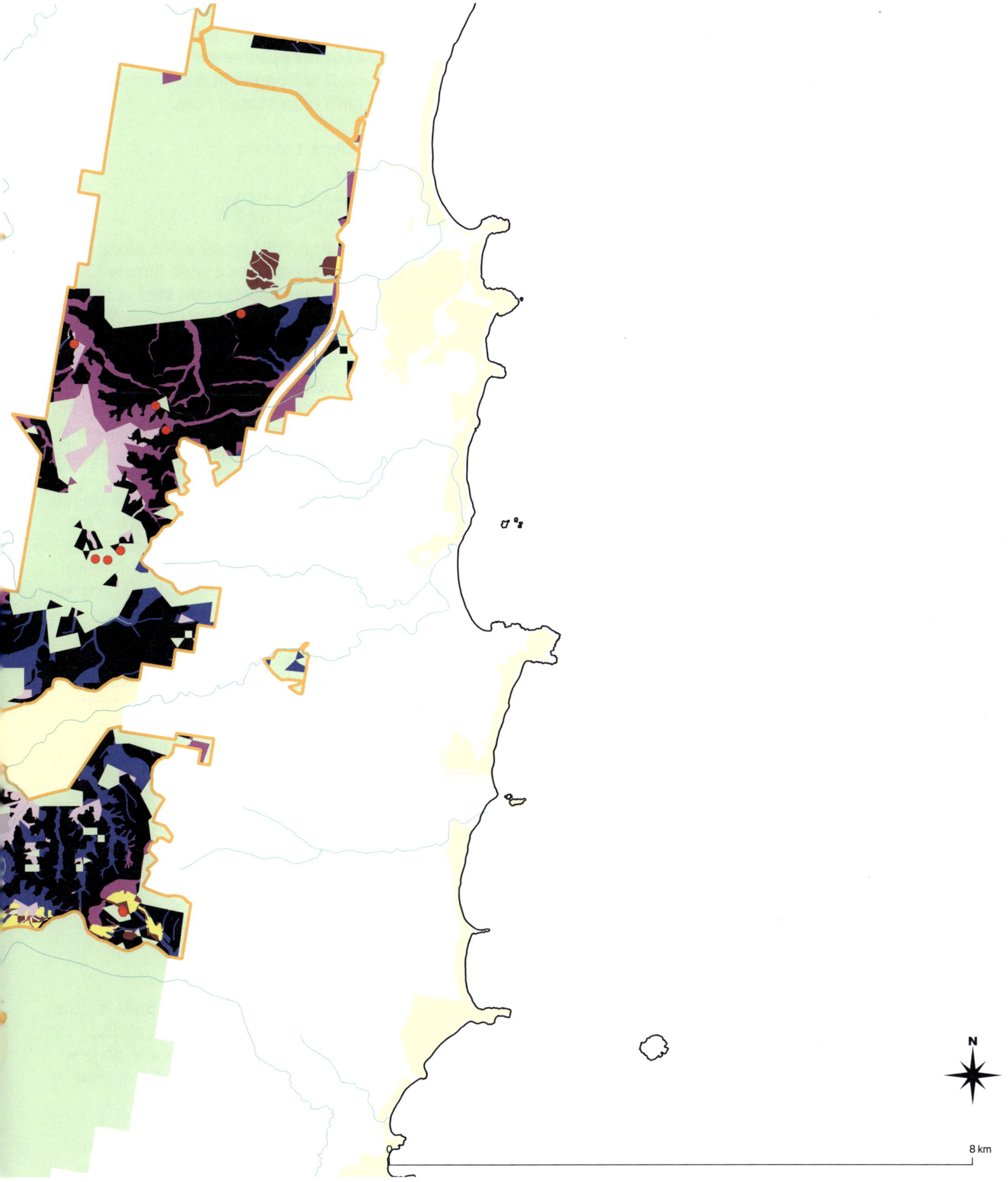

FRIENDS OF CONGLOMERATE
RESERVE DETAILS

FOREST HABITATS

These forests are richly diverse with numerous vegetation communities supporting a range of vegetation types including heath, wetland and aquatic, wet and dry sclerophyll forests, dry rainforest, subtropical rainforest, old-growth, subtropical lowland rainforest, subtropical coastal floodplain forest, rainforest and lowland rainforest on floodplain (Endangered Ecological Communities) and uncommon Montane species inclusions.

Koala habitat within Conglomerate and Wedding Bells grows on poorer sandstone derived soils, nevertheless there is good species variety.

Table 6b. FoC - Koala food trees present and regional use rate.

Tallowwood *Eucalyptus microcorys*	High
Small-fruited Grey Gum *Eucalyptus propinqua*	High
Sandstone Mahognay *Eucalyptus psammitica*	High
Forest Red Gum *Eucalyptus tereticornis*	High
Red Mahogany *Eucalyptus resinifera*	High
Sydney Blue Gum *Eucalyptus saligna*	Significant
Flooded Gum *Eucalyptus grandis*	Significant
White Mahogany *Eucalyptus acmenoides*	Significant
Grey Ironbark *Eucalyptus siderophloia*	Significant
Turpentine *Syncarpia glomulifera*	Significant
Forest Oak *Allocasuarina torulosa*	Significant
Needlebark Stringybark *Eucalyptus planchoniana*	Occasional
Brush Box Lophostemon confertus	Occasional
Broadleaf Paperbark *Melaleuca quinquinervia*	Occasional
Rough-barked Apple *Angophora flopribunda*	Occasional
Smooth-barked Apple *Angophora costata*	Occasional
Pink Bloodwood *Corymbia intermedia*	Occasional
Red Bloodwood *Corymbia gummifera*	Occasional

THREATENED SPECIES

Conglomerate forest contains core habitat for the Brush-tailed Rock Wallaby. It is a regular feeding place for the Glossy Black-Cockatoo and there is a high possibility that the species nests in the central area. Greater Gliders and Yellow-bellied Gliders are known to reside within the boundaries and numerous species of forest microbats inhabit the various suitable niches available. Two particular flagship threatened species here are the Orara Boronia and the Coastal Petaltail Dragonfly which breeds in stationary water bodies which are a feature of the landscapes.

Wedding Bells State Forest contains some Coastal Lowland Subtropical Rainforest (an Endangered Ecological Community) and most of the species are found in neighbouring Conglomerate. Red Bopplenut and the rare Giant Climbing Orchid are both represented here.

See Table 6c. FoC - Threatened species

REASONS FOR INCLUSION

Conglomerate and Wedding Bells forest areas along with Sherwood Nature Reserve form a wide forested corridor linking the Glenreagh escarpment east to the coast and south to the Orara East Koala sub-population. Historical and contemporary records confirm the continued presence of Koalas across this landscape and the suitability of the natural landscape here as habitat for Koalas. There is also an impressive number of threatened flora and fauna to be found here - one third of Australia's microbat species inhabit the Corindi River system which traverses both forests and flows to the Solitary Islands Marine Park and possibly the widest variety of threatened species is to be found throughout this location. They are part of a biologically diverse region supporting many irreplaceable biodiversity assets and Indigenous Cultural Heritage features reflecting past and present relationships with, and within, the landscape.

The proposed reserve contains viable habitat for Koala population and territory expansion which is necessary for their survival. This is a biologically diverse region harbouring a wide variety of threatened species and habitats, and many irreplaceable biodiversity assets, as well as places of important cultural significance reflecting past and present relationships with the landscape. There are cool rainforest riparian corridors and permanent running water, and the area is included in the East Coast Coastal Climate Change Refuge Corridor which is crucial for the continued survival of species, threatened and otherwise.

LANDSCAPE

The landscape of Conglomerate and Wedding Bells features rugged sandstone escarpments, coastal ridges, cliffs, rocky outcrops and overhangs, slopes, gullies and gorges. Elevation ranges from 50mASL to 450mASL. Geological and topographical variances provide a range of niche habitats that support a wide range of associated flora and fauna. Permanent clean water and riparian rainforest corridors provide refuge from the new climate extremes.

Orara Boronia

Perched swamps, wetlands, waterfalls, seeps and soaks, rock overhangs and caves all provide specialised habitat. Numerous springs, aquifers and small streams make up the headwaters and feeder systems for the Corindi River, Woolgoolga Creek and Moonee Creek catchment, and for Sherwood Creek, an important tributary of the Orara River, a major water supply system in the Coffs-Clarence region and a traditional stronghold of the endangered Eastern Freshwater Cod.

KOALAS

These forests have been known as “the home of the Tallowwood”, the tree favoured by Koalas for food here. With multiple altitudes, aspects, water sources, suitable species variety, cool riparian corridors and proximity to the coast, Wedding Bells and Conglomerate provide diverse habitat and protection and relief from heat stress. Records of Koalas in Conglomerate and Wedding Bells are scant but their presence here is undeniable. The current population has low resilience due to ongoing logging operations, the major threat to their continued survival, however the diverse landscape and vegetation provide the range of habitat features necessary to sustain Koalas, and other threatened species including:

- Glossy Black-Cockatoo (Conglomerate forest is regarded as prime breeding habitat)
- Brush-tailed Rock Wallaby (the Escarpment and surrounds are Core Breeding Habitat)
- Greater Glider
- Many endangered, rare and unique flora species including: Orara Boronia, Moonee Quassia and an isolated and rare record of Red Bopplenut of which very few natural stands remain.

CULTURAL HERITAGE

This Escarpment-to-Coast landscape is of deep cultural significance to the north Gumbaynggirr people, with songlines, dreamings, culture and story all embedded in the landscape and its associated flora and fauna.

Evidence abounds in oral and written history, and in physical features and placenames, of a unique and abundant landscape supporting and being supported by the most ancient culture on Earth.

With thanks to Aunty Deb Dootson.

TOURISM AND RECREATION

A wide range of tourism opportunities are evident throughout this landscape. This is further increased with the recent inclusion of Gumbaynggirr language and culture as an immersive tourist experience.

Remnants of petrified forest bear testimony to the age and history of the landscape and there is a wealth of attributes for research and discovery.

The range of species, habitat, vegetation types, and topographical features will provide ongoing opportunity for research and education, and many opportunities will present for the local community and the wider, even global, public to enjoy the amazingly diverse forests in this section of the proposed reserve.

Botany, research, cultural immersion, ecology, geology, recreational fishing and other nature based activities are just some examples, and of course the possibility of seeing Koalas in the wild.

FRIENDS OF CONGLOMERATE
RESERVE DETAILS

WORLD HERITAGE VALUES

Dramatic landscape variations on sandstone sediments with conglomerate and sandstone outcrops in a zone of moderately high winter temperature variations and seasonal heavy flooding with consequent erosion of waterways and escarpments contributes to ongoing change in habitats and species complexes within localised ecosystems.

Gondwanan flora and fauna are unique in the world and the variety evident in these complex and biodiverse forests warrants national and global recognition and protection. The Koala is globally symbolic of this uniqueness, as are the archaeological records and landscape features remaining from, and with, the oldest culture in the world. First Nations-led management of this World Heritage Area will bring unique opportunities in employment and representation.

CURRENT ISSUES

The major threats to Koalas are habitat loss and destruction by logging, land clearing, intense prescribed burns and wildfire. The replacement of mixed native forests with single species plantations is rapidly depleting the habitat Koalas and other threatened species need for survival. Habitat loss is compounded by the intensity and frequency of harvesting and burning. Fragmentation of forests and waterways and loss of connective corridors further push Koalas and other species toward extinction. The continuing depredation of these forested landscapes, with associated loss of species, requires immediate intervention and urgent focus to reverse the accelerating disappearance of these diverse forests and the unique species that depend on them.

Fungal fruiting bodies

FUTURE VISION

Healthy, resilient, diverse and connected forests will ensure the Koalas' survival. Under the permanent protection of a World Heritage listing these forests will provide excellent habitat for territory expansion, population recovery and growth, and the genetic diversity critical to the Koalas (and other threatened species) survival now and into the future, and reverse the current alarming trend of deteriorating health and rapid decline in numbers. With a management regime of custodianship instead of resource extraction, restoration and recovery of degraded habitat and fragmented landscapes are achievable, with diversity, scope, connectivity and permanence being the keys to success.

Un-mapped un-protected old-growth Blue Gum

THREATENED SPECIES

Table 6c. FoC - Threatened species

COMMON NAME	SCIENTIFIC NAME	COMMONWEALTH EPBC ACT STATUS	NSW BC ACT STATUS
FAUNA			
MAMMALS			
Koala	*Phascolarctos cinereus*	Comm End	NSW End
Spotted-tailed Quoll	*Dasyurus maculatus*		NSW Vul
Rufous Bettong	*Aepyprymnus rufescens*		NSW Vul
Brush-tailed Rock Wallaby	*Petrogale penicillata*	Comm Vul	NSW End
Long-nosed Potoroo	*Potorous tridactylus*	Comm Vul	NSW Vul
Greater Glider	*Petauroides volans*	Comm End	NSW End
Yellow-bellied Glider	*Petaurus australis*	Comm Vul	NSW Vul
Squirrel Glider	*Petaurus norfolcensis*		NSW Vul
Brush-tailed Phascogale	*Phascogale tapoatafa*		NSW Vul
Eastern Chestnut Mouse	*Pseudomys gracilicaudatus*		NSW Vul
Common Blossom Bat	*Syconycteris australis* -	NSW Vul	
Large-eared Pied Bat	*Chalinolobus dwyeri*	Comm Vul	NSW Vul
Eastern False Pipistrelle	*Falsistrellus tasmaniensis*		NSW Vul
Golden-tipped Bat	*Kerivoula papuensis*		NSW Vul
Eastern Long-eared Bat	*Nyctophylus corbeni*	Comm Vul	
Little Bent-wing Bat	*Miniopterus australis*		NSW Vul
Large Bent-wing Bat	*Miniopterus orianae oceanensis*		NSW Vul
Hoary Wattled Bat	*Chalinobolus nigrogriseus*		NSW Vul
Southern Myotis	*Myotis macropus*		NSW Vul
Grey-headed Flying Fox	*Pteropus poliocephalus*	Comm Vul	NSW Vul
Greater Broad-nosed Bat	*Scoteanax rueppellii*		NSW Vul
INSECTS			
Coastal Petaltail Dragonfly	*Petalura litorea*		NSW End
FISH (INDICATIVE MAPPING)			
Eastern Freshwater Cod	*Maccullochella ikei*	Comm End	NSW End
Purple-spotted Gudgeon	*Mogurnda adspersa*	Comm End	NSW End
AMPHIBIANS & REPTILES			
Giant Barred Frog	*Mixophyes iteratus*	Comm End	NSW End
Stuttering Frog	*Mixophyes balbus*	Comm Vul	NSW End
Sphagnum Frog	*Philoria sphagnicolus*	Comm Vul	NSW Vul
Stephens' Banded Snake	*Hoplocephalus stephensii*		NSW Vul
BIRDS			
Regent Honeyeater	*Anthochaera phrygia*	Comm Crit End	NSW Crit End
Little Lorikeet	*Glossopsitta pusilla*		NSW Vul
Bush Stone Curlew	*Burhinus grallarius*		NSW End
Glossy Black-Cockatoo	*Calyptorhynchus lathami*	Comm Vul	NSW Vul
Brown Treecreeper	*Climacteris picumnus victoriae*		NSW Vul
Black-necked Stork	*Ephippiorhynchus asiaticus*		NSW End
White-bellied Sea Eagle	*Haliaeetus leucogaster*		NSW Vul
Square-tailed Kite	*Lophoictinia isura*		NSW Vul
Barking Owl	*Ninox connivens*		NSW Vul
Powerful Owl	*Ninox strenua*		NSW Vul
Barred Cuckoo-shrike	*Coracina lineata*		NSW Vul
Scarlet Robin	*Petroica boodang*		NSW Vul

Grey-crowned Babbler	*Pomatostomus temporalis temporalis*		NSW Vul
Wompoo Fruit-dove	*Ptilinopus magnificus*		NSW Vul
Rose-crowned Fruit-dove	*Ptilinopus regina*		NSW Vul
Masked Owl	*Tyto novaehollandiae*		NSW Vul

RARE AND UNIQUE FAUNA			
Superb (Prince Edwards') Lyrebird	*Menura novaehollandiae edwardi* (Not yet threatened but at the northernmost extent of its range here - Dr. G.Clancy)		
FLORA			
VINES			
Milky Silkpod	*Parsonsia dorrigoensis*	Comm End	NSW Vul
Slender Marsdenia	*Marsdenia longiloba*	Comm Vul	NSW End
SHRUBS			
Orara Boronia	*Boronia umbellata*	Comm Vul	NSW Vul
Dwarf Heath Casuarina	*Casuarina defungens*	Comm End	NSW End
	Homoranthus floydii		NSW Vul
Hairy Melichrus	*Melichrus hirsutus*		NSW End
Philotheca	*Philotheca papillata*		NSW Vul
Scrub Turpentine	*Rhodamnia rubescens*	Comm Crit End	NSW Crit End
Moonee Quassia	*Quassia spp Moonee Creek*	Comm End	NSW End
Native Guava	*Rhodomyrtus psidioides*	Comm Crit End	NSW Crit End
ORCHIDS, BULBS, FERNS			
Red-flowered King of the Fairies	*Oberonia Titania*		NSW Vul
Needle-leaf Fern	*Belvisia mucronata*		NSW End
Narrow-leaf Finger Fern	*Grammitis stenophylla*		NSW End
Slender Screw Fern	*Lindsaea incisa*		NSW End
Stink Lily	*Typhonium* sp. aff. *brownii*		NSW End
TREES			
Sandstone Rough-barked Apple	*Angophora robur*	Comm Vul	NSW Vul
Square-fruited Ironbark	*Eucalyptus tetrapleura*	Comm Vul	NSW Vul
Red Bopplenut	*Hicksbeachia pinnatifolia*	Comm Vul	NSW Vul
Rusty Plum	*Niemeyera whitei*		NSW Vul
THREATENED ECOLOGICAL COMMUNITIES			
Lowland Rainforest of Subtropical Australia		Comm Crit End	
Lowland Rainforest on floodplain in the NSW North Coast Bioregion			EEC
Lowland Rainforest in the NSW North Coast and Sydney Basin Bioregions			EEC
Subtropical Coastal Floodplain Forest of the New South Wales North Coast Bioregion			EEC
Coastal Swamp Sclerophyll Forest of New South Wales and South East Queensland		Comm End	
RARE AND UNIQUE PLANTS			
Silver Aspen	*Acronychia wilcoxiana* (western and southern edges of its range in CON SF)		
Steel Box	*Eucalyptus rummeryi*		
Red-leaf Daisy	*Olearia stilwelliae*		
Giant Climbing Orchid.	*Pseudovanilla foliata*		
PROTECTED FOREST/WATER			
High Conservation Value Old Growth			
Rare Forest Type 31			
Class 1 and Class 2 Aquatic Habitats			

To secure permanent protection for all preferred Koala habitats and enhance Koala survival by improving and expanding landscape connectivity, and restoring resilience across disturbed forest landscapes

FRIENDS OF DORRIGO KOALAS

VALUES STATEMENT

PHILOSOPHY

The Friends of Dorrigo Koalas (FoDK) is a community based group of landholders and forest neighbours representing Koala populations on public lands north and northwest of Dorrigo within the Clarence, Bellingen and Coffs Harbour LGAs.

MISSION

To protect endangered Koala populations identified in state forests within the Nymboida River catchment.

To halt and reverse the Koalas current slide to extinction by preventing habitat loss and managing forests to resist climate change and extreme bushfire impacts.

To secure permanent protection for all preferred Koala habitats and enhance Koala survival by improving and expanding landscape connectivity, and restoring fire resilience across disturbed forest landscapes.

Reverse the Koala's status as an endangered species at both the Federal and State level.

GOALS

- Publicly promote the importance of these forests with their resident wildlife populations, and their critical importance to the regional water supply
- Promote the FoDK proposal which aims to improve community understanding of the importance of Dorrigo's Koala habitat forests and to inform and engage stakeholders in Koala conservation
- To increase local support for the inclusion of the FoDK proposal in a Greater Koala Park World Heritage Area
- Survey, record and monitor Koala and glider populations and map threatened species habitats.

Glossy Black-Cockatoo

Koala, Clouds Creek

FRIENDS OF DORRIGO KOALAS
RESERVE DETAILS

Table 7a. FoDK - Reserve data.

Name of Group:	**FRIENDS OF DORRIGO KOALAS**
Total Recorded Koala Sightings	**1,564**
Total area of reserve (ha)	**53,585**
State Forest	**53,585**
State Forests	**CLOUDS CREEK, ELLIS, KANGAROO RIVER, MARENGO, MOONPAR, SHEAS NOB, WILD CATTLE CREEK**
Area zoned Plantation (ha)	**6,145**
Area zoned hardwood plantation	**4,060**
Softwood plantation	**888**
High conservation values (ha)	**14,471**
Rainforest	**6,836**
Old-growth	**2,643**
Disturbed old-growth	**4,992**
Total disturbed HCV since 1998 (ha)	**10,836**

Indicative assessment area of restoration (ha): 43,000
Much of the proposal area is fire impacted with ecosystems in urgent need of regeneration and control of invasive species. Intensity of fire impact varies across the reserve, and some minimally-affected areas have been logged under the guise of salvage. Recent clear-fall forestry in areas zoned plantation has affected several hundred hectares. Detailed research is required to determine full extent of restoration footprint.

Type of Restoration: Remoteness, accessibility and terrain will contribute to higher end costings. Remediation works would need careful assessment and planning to ensure they are efficient and effective considering the natural obstacles to any proposed undertakings. Savings can be made by engaging volunteers and sourcing donations of materials, plant, expertise and grant funding. Detailed research is also required to determine type of restoration.

FRIENDS OF DORRIGO KOALAS
PROPOSED RESERVE

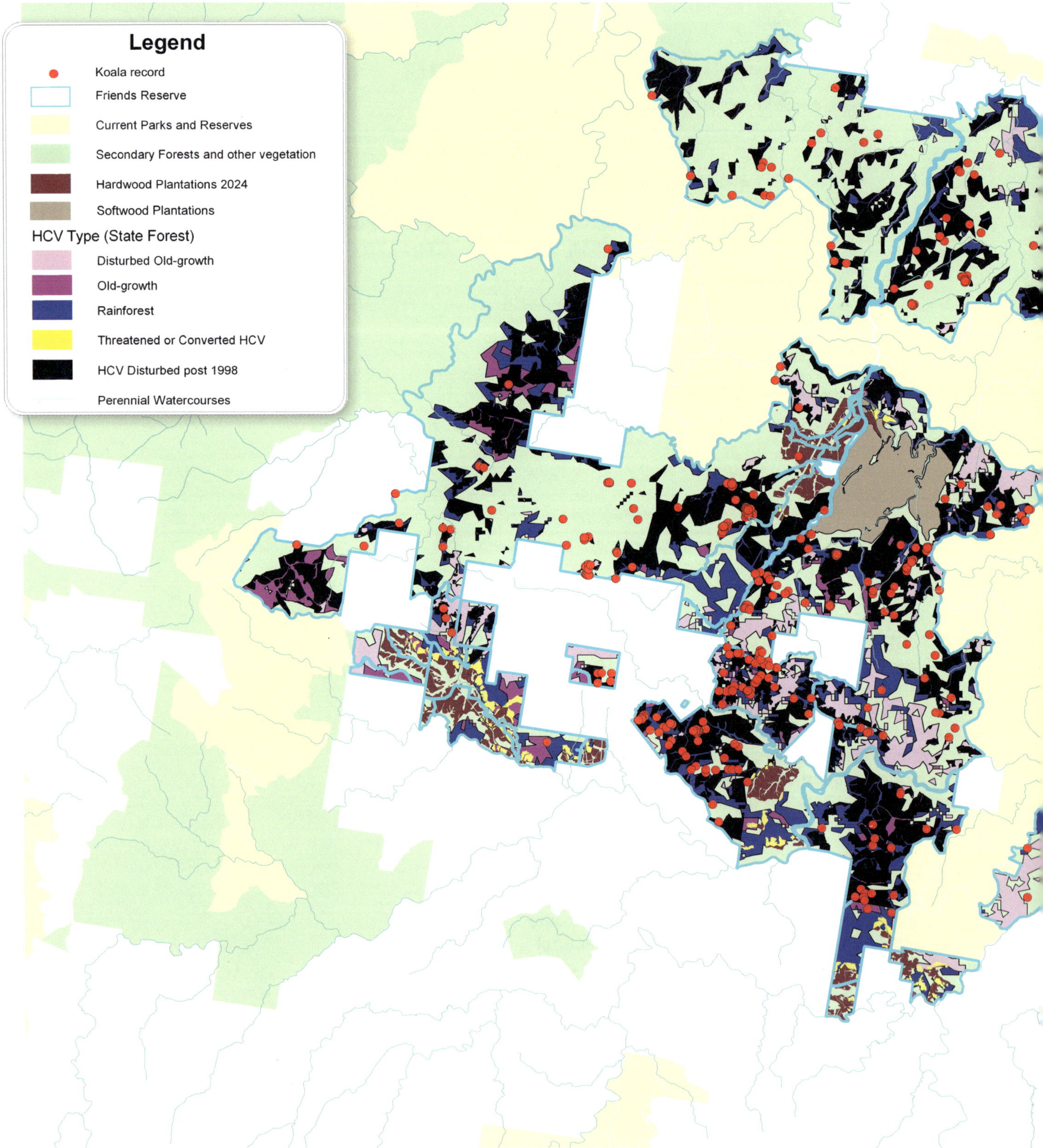

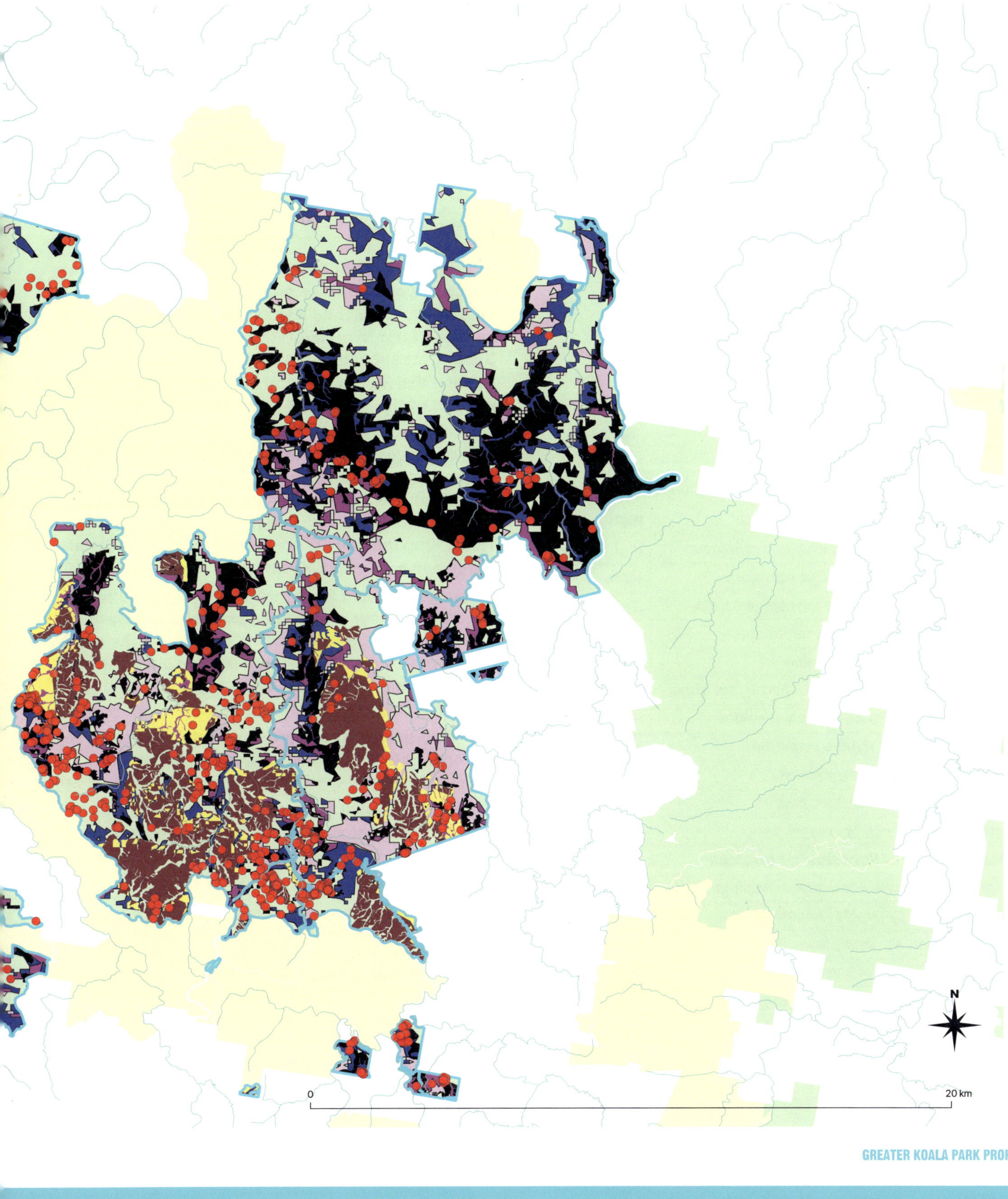
N
0
20 km

FRIENDS OF DORRIGO KOALAS
RESERVE DETAILS

FOREST HABITATS

The scope and terrain of the reserve area translates into a rich variety of well-watered forest types and habitats:

Clarence Dry Sclerophyll Forests,
Cool Temperate Rainforests,
Dry Rainforests,
Eastern Riverine Forests,
Littoral Rainforests,
Montane Bogs and Fens,
Montane Wet Sclerophyll Forests,
New England Dry Sclerophyll Forests,
New England Grassy Woodlands,
North Coast Dry Sclerophyll Forests,
North Coast Wet Sclerophyll Forests,
Northern Escarpment Dry Sclerophyll Forests,
Northern Escarpment Wet Sclerophyll Forests,
Northern Gorge Dry Sclerophyll Forests,
Northern Hinterland Wet Sclerophyll Forests,
Northern Montane Heaths,
Northern Tableland Wet Sclerophyll Forests,
Northern Warm Temperate Rainforests,
Subtropical Rainforests.

Table 7b. FoDK - Koala food trees present and regional use rate.

Tallowwood *Eucalyptus microcorys*	High
Small-fruited Grey Gum *Eucalyptus propinqua*	High
Swamp Mahogany *Eucalyptus robusta*	High
Large-fruited Grey Gum *Eucalyptus punctata*	High
Forest Red Gum *Eucalyptus tereticornis*	High
Grey Box *Eucalyptus moluccana*	High
Narrow-leaved Black Peppermint *Eucalyptus nicholii*	Significant
White Mahogany *Eucalyptus acmenoides*	Significant
Red Mahogany *Eucalyptus resinifera*	Significant
Sydney Blue Gum *Eucalyptus saligna*	Significant
Flooded Gum *Eucalyptus grandis*	Significant
Spotted Gum *Eucalyptus maculata*	Significant
Forest She-oak *Allocasuarina torulosa*	Significant
Silvertop Stringybark *Eucalyptus laevopinea*	Occasional
Red Bloodwood *Corymbia gummifera*	Occasional
Pink Bloodwood *Corymbia intermedia*	Occasional
New England Blackbutt *Eucalyptus andrewsii*	Occasional
Dorrigo White Gum *Eucalyptus dorrigoensis*	Occasional
Smooth-barked Apple *Angophora costata*	Occasional
Brush Box *Lophostemon confertus*	Occasional
Paperbark *Melaleuca*	Occasional
Bottlebrush *Callistemon viminalis*	Occasional

THREATENED SPECIES

The FoDK forests historically support high density, breeding Koala populations, however production forestry regimes disrupt documented Koala territories and fragment known habitats of an array of threatened species.

See Table 7c. FoDK - Threatened species

REASONS FOR INCLUSION

The FoDK Proposal:

- includes over 50,000 hectares of State Forest within the Bellingen, Clarence and Coffs Harbour Shires: Wild Cattle Creek and Kangaroo River State Forests lie east of the Nymboida River (in the Bielsdown & Bobo River subcatchments); and Moonpar, Clouds Creek, Ellis, Sheas Nob and Marengo State Forests lie west of the Nymboida River (in the Blicks River and Clouds Creek subcatchments).
- acknowledges two identified Koala metapopulations on the Mid North Coast of New South Wales – the Coffs Harbour- Guy Fawkes metapopulation, and the Bellinger-Nambucca-Macleay metapopulation. This metapopulation is considered to be of National significance as a core Koala area.
- Protects significant Koala habitat corridors within the Coffs Harbour, north Bellingen and south-western Clarence Valley LGAs, extending inland from the coastal forests at Coffs Harbour, where the Bongil Bongil National Park Koala population is acknowledged as an Area of Intergenerational Significance.
- The hinterland and escarpment forests of Bindarri National Park provide connectivity to the northern tablelands beyond the Guy Fawkes River National Park.
- The FoDK proposal incorporates diverse eucalypt forests in the hinterland and ranges, identified as med-high modelled Koala habitat which also provides significant refugia for a range of threatened fauna and flora species.

The Dorrigo Plateau's *eucalyptus* species diversity is recognised internationally for inclusion to World Heritage status, with the Dorrigo Koala reserve area linking coastal Koala habitat through Bindarri NP where FoDK state forests improve the east-west corridor, alongside Cascade NP and Nymboi-Binderay NP to connect the western forests through Chaelundi wilderness and Guy Fawkes River NP to known Koala populations on the Northern Tablelands.

Clouds Creek

Protection of identified Koala and threatened species populations within the reserve forests will enhance habitat connectivity and improve the health and survival of Koala and glider populations recovering from recent bushfire impacts.

The cumulative impacts to water security from industrial logging in public forests, drought, extreme bushfire and rainfall events has negatively affected the quality of the regional water supply across the Nymboida River catchments. Conservation and restoration of Dorrigo's catchment forests will support climate mitigation and the impacts of future extreme weather events predicted to cause catchment-wide downstream consequences for coastal communities.

LANDSCAPE

The tall, wet forests of the western Dorrigo Plateau provide critical landscape connectivity for endangered and vulnerable wildlife, vital for climate adaptation. Linking the remnant wet coastal hinterland forests to a rich wilderness of higher elevation subtropical, warm temperate and cool temperate rainforests with areas of remnant high conservation value old-growth, in permanently protected existing national parks and state forest reserves will enhance the national reserve system and the regional water supply.

These high altitude landscapes provide climate refuge and support tall wet forests that are climate moderating influencers, as well as montane heaths and dry forests, ephemeral and permanent water supplies, gorges a nd grasslands - each supporting a diversity of contingent species.

KOALAS

Koala habitat in the FoDK reserve area has been subject to ongoing fragmentation. A range of geological, topographical and vegetative features support diverse eucalypt species that are preferred Koala food trees, some being specific to particular areas. Linking these isolated habitat patches is necessary for the ongoing survival of both Koalas and the food trees they rely on. Permanent water and higher altitudes provide favourable conditions for Koalas as reflected by their significant, persistent presence in the face of, and despite continued habitat loss.

CULTURAL HERITAGE

FoDK recognises the cultural significance of Country within the reserve proposal, and supports its protection and culturally appropriate management.

European heritage relates to land clearing for grazing and the timber and mining industries and includes physical remains such as timber tramlines, bridges and logging arches.

TOURISM, RECREATION, CULTURAL LEARNING

The Dorrigo Plateau has a reputation for spectacular scenery and forests. Breathtaking tall wet forests and rainforests contain many cascades and waterfalls that feed into rivers which provide opportunities for recreational fishing and kayaking, wilderness walks, botanical and medical research, scientific and educational studies.

Jaliigirr Biodiversity Alliance in partnership with the Centre for Ecological Learning delivered a pilot project (Traditional Cultural Corridors) which identified Gliders and glideways as highly important indicators of connectivity health, offering "a recognisable and iconic way for the broader community

Blue Gum, Clouds Creek

RESERVE DETAILS

to become involved". Work has continued on Landscape Connections in the Great Eastern Ranges corridor to connect communities and culture through their Corridors Project.

WORLD HERITAGE VALUES

EUCALYPTUS DIVERSITY

Australia's eucalypts have been identified as being of outstanding universal value, and the Koala as an outstanding example of a vertebrate species with an evolutionary dependence on eucalypts. The FoDK reserve forests support a range of natural values including core Koala and Glider habitat, and old-growth forest, and are part of an extensive belt of forest covering the eastern fall of Dorrigo Plateau.

Superb specimens of Araucaria (the most primitive of the world's conifers) and Nothofagus are found in these forests, along with Lyrebirds, Rufous Scrub bird, Bowerbirds, and Pouched Frogs. These relict Gondwanan flora and fauna provide a living link with Australia's evolution.

The Demon Fault and Ebor Volcano are of geoheritage significance through the resultant mosaic of disjunct and intermingled geological influences on the landscapes, soils, flora and fauna, and as the Great Divide erodes the Great Escarpment continues to make a westerly retreat along the northeastern edge of the Dorrigo Plateau.
The wild and scenic, nationally significant Nymboida River runs through these parks. The river is a major water supply for the Clarence–Coffs region, supporting the only stable wild-breeding population of the endangered eastern freshwater cod in the world, and is used for commercial and recreational whitewater rafting.

CURRENT ISSUES

NSW Koalas are at risk of extinction and listed as an endangered species at both the Federal and State level due to habitat loss, climate change and extreme bushfire impacts. Ongoing industrial logging is impacting remaining habitat values in fire-impacted refugia forests.

Repercussions of this land clearing are loss of hollow-bearing trees, arboreal mammal and threatened Glider habitat (Greater Glider & Yellow Bellied Glider habitats) and sedimentation of the Clarence-Coffs Harbour regional water supply requiring costly filtration and further threatening aquatic species and water security.

FUTURE VISION

The proposed GKPWHA will include the protection of 200,497 hectares of public eucalypt forests and incorporate existing National Parks into a Koala sanctuary (of nearly 492,000 hectares) identified as having a range of outstanding universal values.

Protection of the forest ecosystems of the Nymboida River catchments will improve water quality and water security for the region, enhancing the biodiversity and fire resilience of forests as they grow old, providing a range of ecological and social benefits for business and recreational pursuits.

THREATENED SPECIES

Table 7c. FoDK - Threatened species

COMMON NAME	SCIENTIFIC NAME	COMMONWEALTH EPBC ACT STATUS	NSW BC ACT STATUS
MAMMALS			
Koala	*Phascolarctos cinereus*	Comm End	NSW End
Rufous Bettong	*Aepyprymnus rufescens*		NSW Vul
Parma Wallaby	*Notamacropus parma*	Comm Vul	NSW Vul
Brush-tailed Rock Wallaby	*Petrogale penicillata*		NSW End
Long-nosed Potoroo	*Potorous tridactylus*		NSW Vul
Red-legged Pademelon	*Thylogale stigmatica*		NSW Vul
Spotted-tailed Quoll	*Dasyurus maculatus*	Comm End	NSW Vul
Brush-tailed Phascogale	*Phascogale tapoatafa*		NSW Vul
Eastern Chestnut Mouse	*Pseudomys gracilicaudatus*		NSW Vul
Hastings River Mouse	*Pseudomys oralis*	Comm End	NSW Vul
Greater Glider	*Petauroides volans*	Comm End	NSW End
Yellow-bellied Glider	*Petaurus australis*	Comm Vul	NSW Vul
Squirrel Glider	*Petaurus norfolcensis*		NSW Vul
Eastern False Pipistrelle	*Falsistrellus tasmaniensis*		NSW Vul
Eastern Coastal Freetail Bat	*Micronomus norfolkensis*		NSW Vul
Little Bent-wing Bat	*Miniopterus australis*		NSW Vul
Large Bent-wing Bat	*Miniopteris orianae oceanensis*		NSW Vul
Southern Myotis	*Myotis macropus*		NSW Vul
Golden-tipped Bat	*Phoniscus papuensis*		NSW Vul
Greater Broad-nosed Bat	*Scoteanax rueppellii*		NSW Vul
Eastern Cave Bat	*Vespadelus troughtonii*		NSW Vul
FISH			
Eastern Freshwater Cod	*Maccullochella ikei*	Comm End	NSW End
Eel-tailed Catfish	*Tandanus tandanus*	Comm Vul	NSW End
AMPHIBIANS			
Pouched Frog	*Assa darlingtonii*		NSW Vul
Stuttering Frog	*Mixophyes balbus*	Comm Vul	NSW End
Giant Barred Frog	*Mixophyes iteratus*	Comm End	NSW End
Sphagnum Frog	*Philoria sphagnicola*	Comm Vul	NSW Vul
BIRDS			
Rufous Scrub Bird	*Atrichornis rufescens*	Comm End	NSW Vul
Sooty Owl	*Tyto tenebricosa*		NSW Vul
Barking Owl	*Ninox connivens*		NSW Vul
Masked Owl	*Tyto novaehollandiae*		NSW Vul
Powerful Owl	*Ninox strenua*		NSW Vul
Little Eagle	*Hieraaetus morphnoides*		NSW Vul
White-bellied Sea Eagle	*Haliaeetus leucogaster*		NSW Vul
Little Lorikeet	*Parvipsitta pusilla*		
Wompoo Fruit-dove	*Ptilinopus magnificus*		
Rose-Crowned Fruit-dove	*Ptilinopus regina*		
Superb Fruit-dove	*Ptilinopus superbus*		
Square-tailed Kite	*Lophoictinia isura*		
Regent Honeyeater	*Anthochaera phrygia*	Comm Crit End	NSW Crit End

Grey-crowned Babbler	*Pomatostomus temporalis*		
Comb-crested Jacana	*Irediparra gallinacea*		NSW Vul
Black-necked Stork	*Ephippiorhynchus asiaticus*		NSW End
Brown Treecreeper (eastern sbsp)	*Climacteris picumnus victoriae*	Comm Vul	NSW Vul
White-throated Needletail	*Hirundapus caudacutus*	Comm Vul	NSW Vul

Pale-headed Snake	*Hoplocephalus bitorquatus*		NSW Vul
Stephens' Banded Snake	*Hoplocephalus stephensii*		NSW Vul
TREES			
Square-fruited Ironbark	*Eucalyptus tetrapleura*	Comm Vul	NSW Vul
Sandstone Rough-barked Apple	*Angophora robur*	Comm Vul	NSW Vul
Woodland Babingtonia	*Babingtonia sylvestris*		NSW End
	Bertya sp. (Clouds Creek, M. fatemi 4)		NSW End
Gympie Stinger	*Dendrocnide moroides*		NSW End
Beadle's Grevillea	*Grevillea beadleana*	Comm End	NSW End
Tall Velvet Seaberry	*Haloragis exalata subsp velutina*		NSW Vul
White Karambal	*Harneria hygrophiloides*		NSW End
	Kardomia prominens	Comm Crit End	NSW Crit End
Dorrigo Daisy Bush	*Olearia flocktoniae*	Comm End	NSW End
Weeping Melaleuca	*Melaleuca irbyana*		NSW End
Johnson's Cycad	*Macrozamia johnsonii*	Comm End	NSW End
Brush Sauropus	*Phyllanthus microcladus*		NSW End
	Pomaderris queenslandica		NSW End
Rainforest Cassia	*Senna acclinis*		NSW End
VINES			
White-flowered Wax Plant	*Cynanchum elegans*	Comm End	NSW End
Slender Marsdenia	*Marsdenia longiloba*	Comm Vul	NSW End
Milky Silkpod	*Parsonsia dorrigoensis*	Comm End	NSW Vul
Cryptic Forest Twiner	*Tylophora woollsii*	Comm End	NSW End
GRASSES, FORBS, ORCHIDS			
Hairy Joint Grass	*Arthraxon hispidus*		NSW Vul
Small Pale Grass Lily	*Caesia parviflora*		NSW End
Narrow Leaf Finger Fern	*Grammitis stenophylla*		NSW End
Brown Butterfly Orchid	*Sarcochilus dilatatus*		NSW End
Ravine Orchid	*Sarcochilus fitzgeraldii*	Comm Vul	NSW Vul
Austral Toadflax	*Thesium australe*	Comm Vul	NSW Vul
ENDANGERED ECOLOGICAL COMMUNITIES			
White Gum Moist Forest in the NSW North Coast Bioregion			EEC
Lowland Rainforest of Subtropical Australia		Comm End	
Grey Box-Grey Gum Wet Sclerophyll Forest in the NSW North Coast Bioregion			EEC
Subtropical Coastal Floodplain Forest of the New South Wales North Coast Bioregion			EEC
Montane Peatlands and Swamps of the New England Tableland, NSW North Coast, Sydney Basin, South East Corner, South Eastern Highlands and Australian Alps bioregions			EEC
RARE AND ENDEMIC			
Dorrigo Plum	*Endiandra introrsa*		
Steel Box	*Eucalyptus rummeryi*		

The reserve should become World Heritage Area due to its outstanding universal value as a pristine and biodiverse Australian ecosystem, providing critical habitat and water sources for numerous endangered species

FRIENDS OF KALANG HEADWATERS

VALUES STATEMENT

PHILOSOPHY

The Friends of Kalang Headwaters is a community-based organisation, including the Bellingen Environment Centre, National Parks Association of NSW, businesses, and individuals from our community.

MISSION

To protect, promote and preserve the cultural and natural values and heritage of the Kalang river forests and headwaters from logging.

GOALS

Achieve permanent protection of the Kalang headwaters and forests by extending the Baalijin Nature Reserve via the Headwaters Conservation Proposal – A nature reserve for the Upper Kalang, Middle Bellinger and Nambucca Rivers, and promoting their inclusion in a wider, permanently protected Greater Koala World Heritage Area. Achieve these goals through non-violent, direct action.

Koala tree, Oakes

Cunjevoi, Oakes

Proposed reserve, Oakes forest, Bellbucca Road

FRIENDS OF KALANG HEADWATERS
RESERVE DETAILS

Table 8a. FoKH - Reserve data.

Name of Group:	**FRIENDS OF KALANG HEADWATERS**
Total Recorded Koala Sightings	**298**
Total area of reserve (ha)	**13,234**
State Forest	**13,234**
State Forests	**BUCKRABENDINNI, DIE HAPPY (PART), IRISHMAN, OAKES, ROSES CREEK, SCOTCHMAN (PART)**
High conservation values (ha)	**6,519**
Rainforest	**2,899**
Old-growth	**2,338**
Disturbed old-growth	**1,282**
Total disturbed HCV since 1998 (ha)	**125**

Indicative assessment area of restoration (ha): 4,500

The reserve has seen little forestry since 1998. Natural recovery and regrowth can be expected to be high. Additonal damage resulting from recent forestry-related roadworks in Oakes will require amelioration. Assessment of the fire footprint and the level of remediation will also be required.

Type of restoration:

Dependent on level of weed infestation and/or site preparation works required, eg machinery may be needed for initial clearing, thinning, ground remediation or preparation works. Terrain and remoteness will add to costs. Annual follow-up maintenance will be necessary (3-10 years). Savings can be made with volunteers contributing labour and by sourcing donations of plants, materials, expertise and grant funding.

FRIENDS OF KALANG HEADWATERS
PROPOSED RESERVE

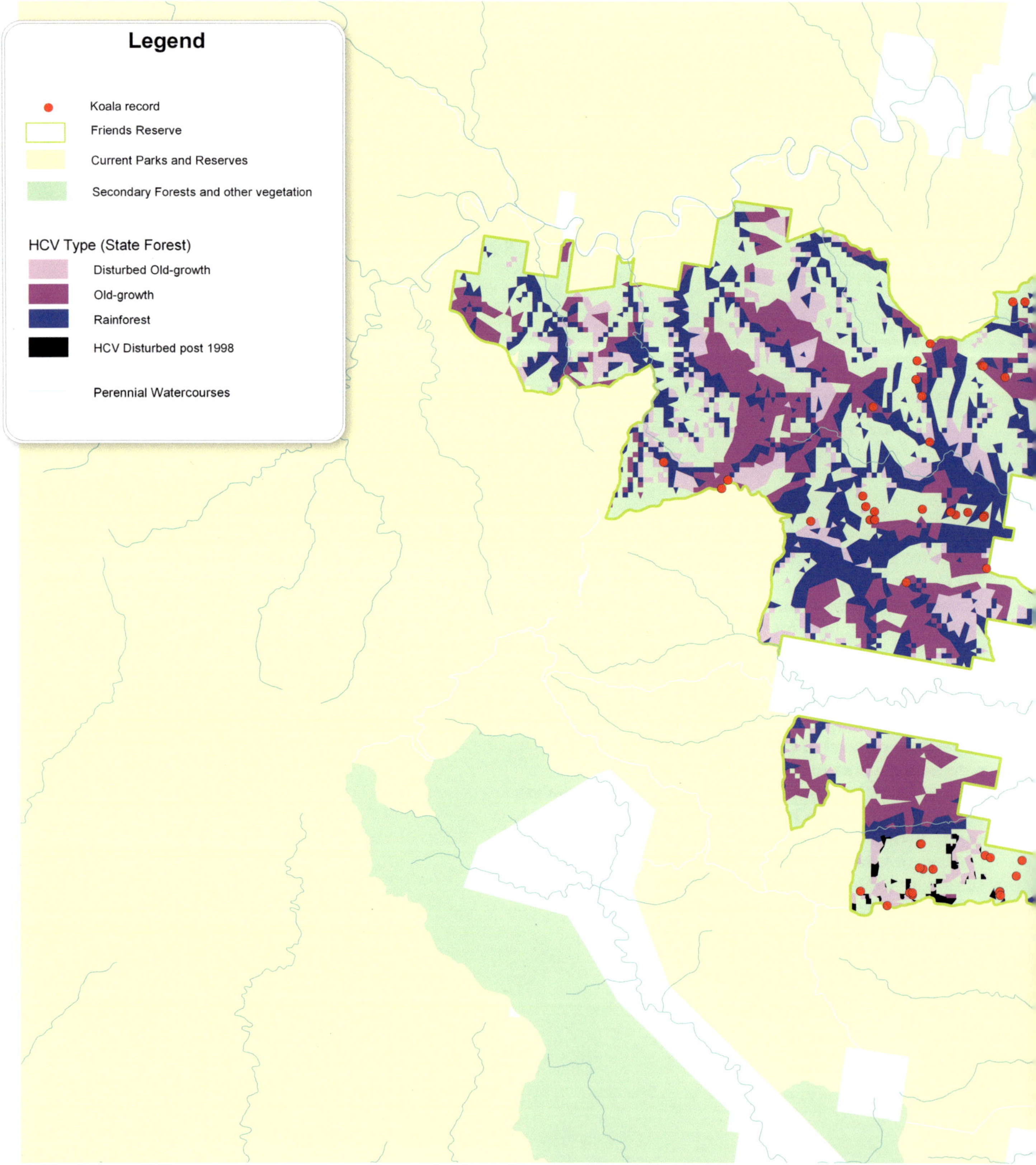

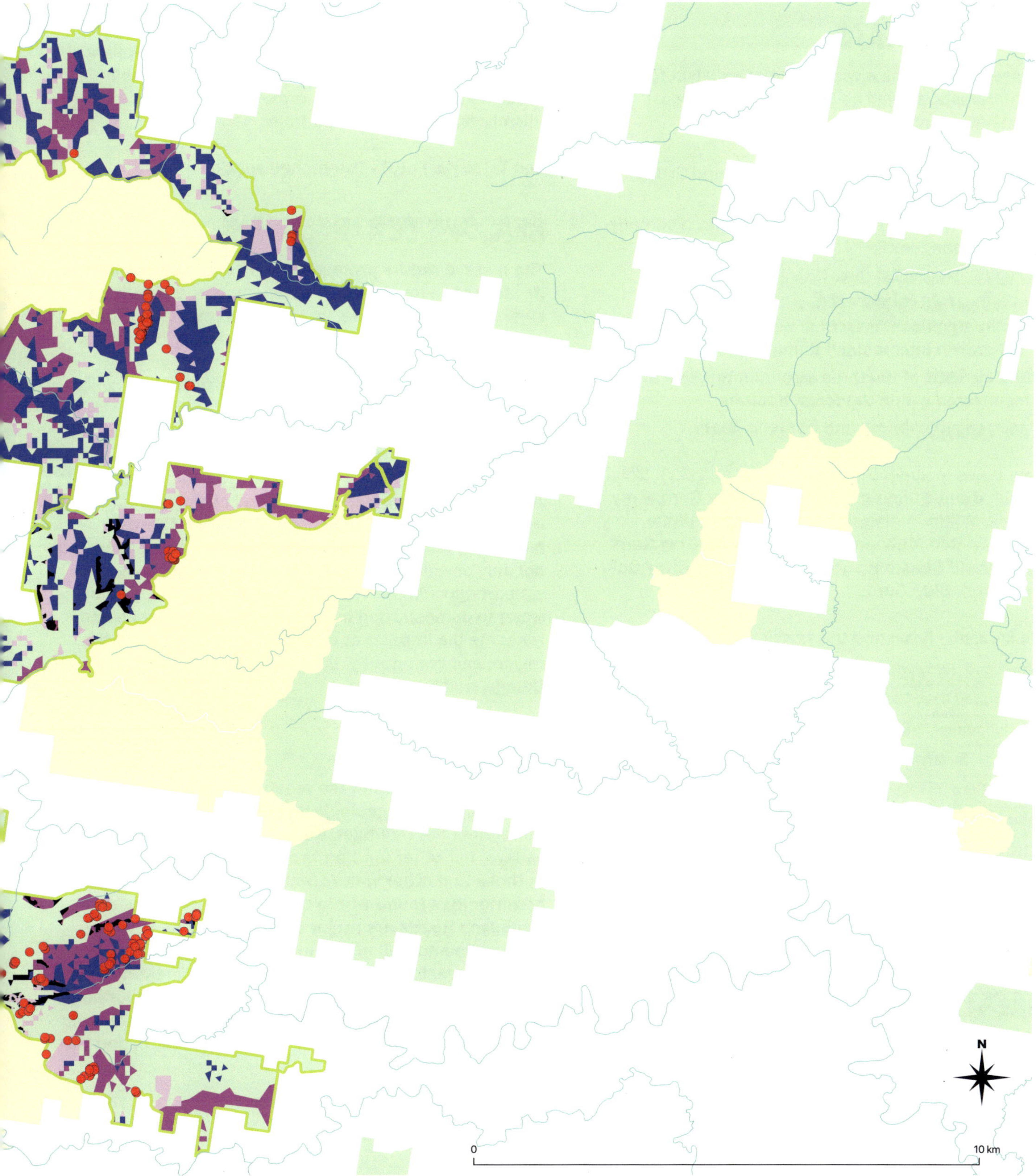
N
0
10 km

RESERVE DETAILS

FOREST HABITATS

The reserve comtains a wide range of eucalypt forests and rainforests, as well as Endangered Ecological Communities, including:

Lowland Rainforest in NSW North Coast and Sydney Basin Bioregion (Comm End);
Lowland Rainforest of Subtropical Australia (Federally Critically Endangered);
Subtropical rainforest (inc. TECs);
Warm temperate rainforest (Inc. TECs);
Nationally significant intact tall wet sclerophyll forests and old-growth interspersed throughout reserve;
Aquatic systems of the three catchments which are interconnected within the forest habitats;
Dry sclerophyll habitats and **Montane Heath**.

Forest types include: Blackbutt (moist and dry); Brush Box; Fig, Giant Stinging Tree, Myrtle; Grey Gum, Grey Ironbark, White Mahogany; Narrow-leaved White Mahogany, Red Mahogany, Grey Ironbark, Grey Gum; New England Blackbutt, Sydney Blue Gum; River Oak; Tallowwood, Blue Gum.

Table 8b. FoKH - Koala food trees present and regional use rate.

Tallowwood *Eucalyptus microcorys*	High
Small-fruited Grey Gum *Eucalyptus propinqua*	High
Red Mahogany *Eucalyptus resinifera*	High
Blue Gum *Eucalyptus saligna*	Significant
Flooded Gum *Eucalyptus grandis*	Significant
White Mahogany *Eucalyptus acmenoidies*	Significant
Coastal Ironbark *Eucalyptus siderophloia*	Significant
Turpentine *Syncarpia glomulifera*	Significant
Brush Box *Lophostemon confertus*	Significant
Forest Oak *Allocasuarina torulosa*	Significant

THREATENED SPECIES

The reserve contains a wide range of threatened species that have adapted to, and are dependent on specific, but diverse, ecosystems. Koala species sightings have been found across the reserve, and unique species including the Rufous Scrub Bird, which harks back to an earlier evolutionary era, abound. Glossy Black-Cockatoo nests have also been found withn the proposed reserve.

Acquatic ecosystems include riverine habitat for the Purple Spotted Gudgeon and Bellinger River Snapping Turtle, both critically endangered. Other threatened species include forest owls, raptors, gliders, doves, microbats, marsupials, and frogs.

See Table 8c. FoKH - Threatened species

REASONS FOR INCLUSION

The reserve merits protection particularly on account of its landscape-level values, which support a bio-diverse suite of flora and fauna, including an important Koala population, but are currently threatened by the area's ongoing status as a wood production zone, which challenges the vision of a sustainable future.

This reserve represents a critical connecting link to all the adjacent formal reserves including the New England and Dorrigo Gondwana World Heritage Areas, the Baalijin and Juugawaarri Nature Reserves, as well as the Gumbaynggirr National Park and State Conservation Area. These headwater forests are the source for three major rivers that not only provide a wide array of ecosystem services. They support agriculture, aquaculture and recreation, provide water to domestic and business users, and regulate and mitagate the impacts of climate change throughout the region, and importantly, the urban populations of Bellingen, Urunga, and Nambucca.

LANDSCAPE

The landscapes within the reserve are outstanding, both visually, and geologically. The Ebor volcano caldera dominates the Bellinger, Kalang and Nambucca River valleys, but other landforms and geologies are found in these and other valleys, and influence the ecological communities found within them. The Horseshoe- and Bellbucca Roads are major connecting routes linking Bowraville and Bellingen, and offer touring parties a range of spectacular vistas and intimate forested views. The steep slopes around the headwaters of the Kalang River, make for a precipitous descent for rapidly-flowing creeks and streams, down to the valley bottom, and the ocean beyond.

Morning mist from Scotchman Peak

KOALAS

The Mid North Coast is home to a number of important meta populations of Koala and make up a large component of the Northern Koala population of NSW and Queensland. While records for Koalas are relatively low given the size of the reserve, it is known that healthy breeding populations are found across the reserve, with more recent investigations demonstrating good numbers in areas such as Roses Creek and Scotchman. Koala feed trees abound, and deep rainforest gullies afford a cool refuge during hot summer months. The nature of the three main headwater regions allows for both a diverse range of habitats, while simultaneously facilitang movement and cross-breeding of otherwise distinct Koala populations, critical for ensuring the diversity of the species, and avoiding disese through inbreeding, common in other more disturbed landscapes.

CULTURAL HERITAGE

FoKH understands there are many cultural values across the proposed reserve. These values will be best maintained through cooperative and protective management strategies, under the direction and leadership of the Gumbaynggirr Nation.

TOURISM AND RECREATION

The proposed reserve will link the coast to the New England Wilderness Trail and the National Trail, with the possibility of making one of the longest wilderness walks in the world. The National trail begins in Victoria at the end of the Black Spur (Toolangi) and runs along the Great Divide into Queensland. Part of the New England Wilderness walk adjoins Oakes with access via Horseshoe Road, and this connection and proximity to the New England Wilderness, Gondwanan Rainforest of Australia (WHA) would make world class trekking with options to start, end, or veer off to the coast. Spectacular scenery and geological features at lookout points and within the forest, and remnants of old-growth rainforests, which remained as refuge during the 2019-2020 wildfires due to their ability to suppress fire, are to be encountered across these forests. These values lend themselves to medical and scientific research, recreational visitation, and opportunities for learning from traditional knowledge.

WORLD HERITAGE VALUES

The reserve should become a World Heritage Area due to its outstanding universal value as a pristine and biodiverse Australian ecosystem, providing critical habitat and water sources for numerous endangered species. Kalang Headwater forests are the source for three major rivers which supply life to many threatened ecosystems, rainforests, ancient vegetation, and iconic species. The natural history of the area, represented by its unique geological features of ancient rock formations and deep gorges, further contribute to its significance as a World Heritage site. This area has a long and continued human history which bears unique testimony to the importance of this exceptional area.

FRIENDS OF KALANG HEADWATERS
RESERVE DETAILS

Un-mapped, un-protected old-growth Brushbox, Oakes

Pyroclast river rock Upper Kalang

CURRENT ISSUES

One of the major current issues facing these forests is active, but not yet implemented, industrial logging practices which would be damaging to the fragile soils on precipitous landscapes, which in turn threatens water quality and security. Environmental groups, the global scientific community, and local residents have been advocating for protection to preserve and sustain the globally significant unique features. Additionally climatic changes such as the increased frequency of fire and changing rainfall patterns further exacerbate the threat of degradation and destruction, particularly on exposed soils and fragmented forests.

FUTURE VISION

Protective and collaborative management of these forests as a reserve will ensure this core Gondwanan refuge remains of global significance for present and future generations, and is acknowledged for its value as critical for the survival of ancient ecosystems and species. Collaborative efforts between the local community, non governmental organisations, government agencies and First Nations people, will ensure the preservation of the incredibly valuable cultural and ecological heritage of the area, while promoting sustainable tourism and education initiatives into the future.

THREATENED SPECIES

Table 8c. Threatened species

COMMON NAME	SCIENTIFIC NAME	COMMONWEALTH EPBC ACT STATUS	NSW BC ACT STATUS
FLORA			
Milky Silkpod	Parsonsia dorrigoensis	Comm Vul	NSW End
Slender Marsdenia	Marsdenia longiloba	Comm Crit End	NSW Crit End
Scrub Turpentine	Rhodamnia rubescens	Comm Crit End	NSW Crit End
Rusty Plum	Neimeyera whitei		NSW Vul
Mountain angelica	Gingidia rupicola	Comm End	NSW End
FAUNA			
REPTILES			
Bellinger River Snapping Turtle	Myuchelys georgesii	Comm Crit End	NSW Crit End
Stephens' Banded Snake	Hoplocephalus stephensii		NSW Vul
AMPHIBIANS			
Sphagnum Frog	Philoria sphagnicolus		NSW Vul
Stuttering Frog	Mixophyes balbus	Comm End	NSW End
Giant Barred Frog	Mixophyes iteratus	Comm Vul	NSW Vul
BIRDS			
Rufous Scrub-bird	Atrichornis rufescens	Comm End	NSW Vul
Wompoo Fruit-dove	Ptilinopus magnificus		NSW Vul
Rose-crowned Fruit-dove	Ptilinopus regina		NSW Vul
Glossy Black-Cockatoo	Calyptorhynchus lathami	Comm Vul	NSW Vul
Powerful Owl	Ninox strenua		NSW Vul
Sooty Owl	Tyto tenebricosa		NSW Vul
Varied Sittella	Daphoenositta chrysoptera		NSW Vul
Little Lorikeet	Glossopsitta pusilla		NSW Vul
Osprey	Pandion cristatus (haliaetus)		NSW Vul
White-bellied Sea Eagle	Haliaeetus leucogaster		NSW Vul
Square-tailed Kite	Lophoictinia isura		NSW Vul
White-throated Needletail	Hirundapus caudacutus	Comm Vul	NSW Vul
MAMMALS			
Koala	Phascolarctos cinereus	Comm End	NSW End
Long-nosed Potoroo	Potorous tridactylus	Comm Vul	NSW Vul
Spotted-tailed Quoll	Dasyurus maculatus	Comm End	NSW Vul
Common Planigale	Planigale maculata		NSW Vul
Red-legged Pademelon	Thylogale stigmatica		NSW Vul
Parma Wallaby	Macropus parma	Comm Vul	NSW Vul
Greater Glider	Petauroides volans	Comm Vul	NSW Vul
Eastern Pygmy Possum	Cercartetus nanus		NSW Vul
Brush-tailed Phascogale	Phascogale tapoatafa		NSW Vul
Golden-tipped Bat	Phoniscus papuensis		NSW Vul
Little Bent-wing Bat	Miniopterus australis		NSW Vul
Greater Broad-nosed Bat	Scoteanax rueppellii		NSW Vul
Grey-headed Flying-fox	Pteropus poliocephalus	Comm Vul	NSW Vul
Hastings River Mouse	Pseudomys oralis	Comm End	NSW End
Yellow Bellied Glider	Petaurus australis	Comm Vul	NSW Vul
FISH			
Purple Spotted Gudgeon	Mogurnda adspersa		NSW Vul (Possibly extinct)

The way people can survive and their interaction with biodiversity in drying climates is important for the world as we enter a new stage of climate change

FRIENDS OF NEWRY

VALUES STATEMENT

PHILOSOPHY

Friends of Newry is a diverse local community group. It is based on cooperation and information exchange, and is made up of forest neighbours and like-minded people, while being apolitical and representing our diverse local community. We collaborate on community projects to research and protect native species within the reserve and surrounding environments.

MISSION

We aim to assist in:

- the preservation of local biodiversity, and the long-term protection of endemic and endangered species. We focus on habitat connectivity and pathways, forest recovery, permanent protection for important habitat, and ecologically sustainable forest management.
- the protection of our community from bush fire and climate change risks, and to support tourism and the health of our community by supporting sustainable forest management.

GOALS

- Ensure Little Newry and Newry retain their integrity as unbroken pathways with other habitats
- Protect endemic and endangered populations by mapping their presence within the reserve
- End logging in habitat of endangered or vulnerable species
- Facilitate communication about sustainable timber, tourism, environmental technology, and the sciences.
- Research and communicate the positive effects forests have on mental health
- Educate our local community about the reserve's biodiversity assets
- Conduct regular citizen science activities to engage our community with their local environment.

Coastal eucalypt forest, Little Newry

Forest giant, Little Newry forest

FRIENDS OF NEWRY
RESERVE DETAILS

Table 9a. FoN - Reserve data.

Name of Group:	**FRIENDS OF NEWRY**
Total Recorded Koala Sightings	**128**
Total area of reserve (ha)	**4,547**
State Forest	**4,547**
State Forests	**NEWRY, LITTLE NEWRY**
Area zoned Plantation (ha)	**723**
Area zoned hardwood plantation	**696**
High conservation values (ha)	**262**
Rainforest	**87**
Old-growth	**133**
Disturbed old-growth	**42**
Total disturbed HCV since 1998 (ha)	**163**

Indicative assessment area of restoration (ha): 1,600

- 489 ha clearfell replant with biodiverse habitat species of Newry and manage weed regrowth.
- 500 ha (approx) converted to Blackbutt monoculture 2019. Thinning existing blackbutt seedlings, replanting with at least 50% biodiverse appropriate tree species
- 600 ha native forest cleared in high value Koala habitat by recent logging replanted and invasive weed management.

Type of restoration: Dependent on level of weed infestation and/or site preparation works required, eg machinery may be needed for initial clearing, thinning, ground remediation or preparation works. Annual follow-up maintenance will be necessary (3-10 years). Savings can be made with volunteers contributing labour and by sourcing donations of plants, materials, expertise and grant funding.

FRIENDS OF NEWRY
PROPOSED RESERVE

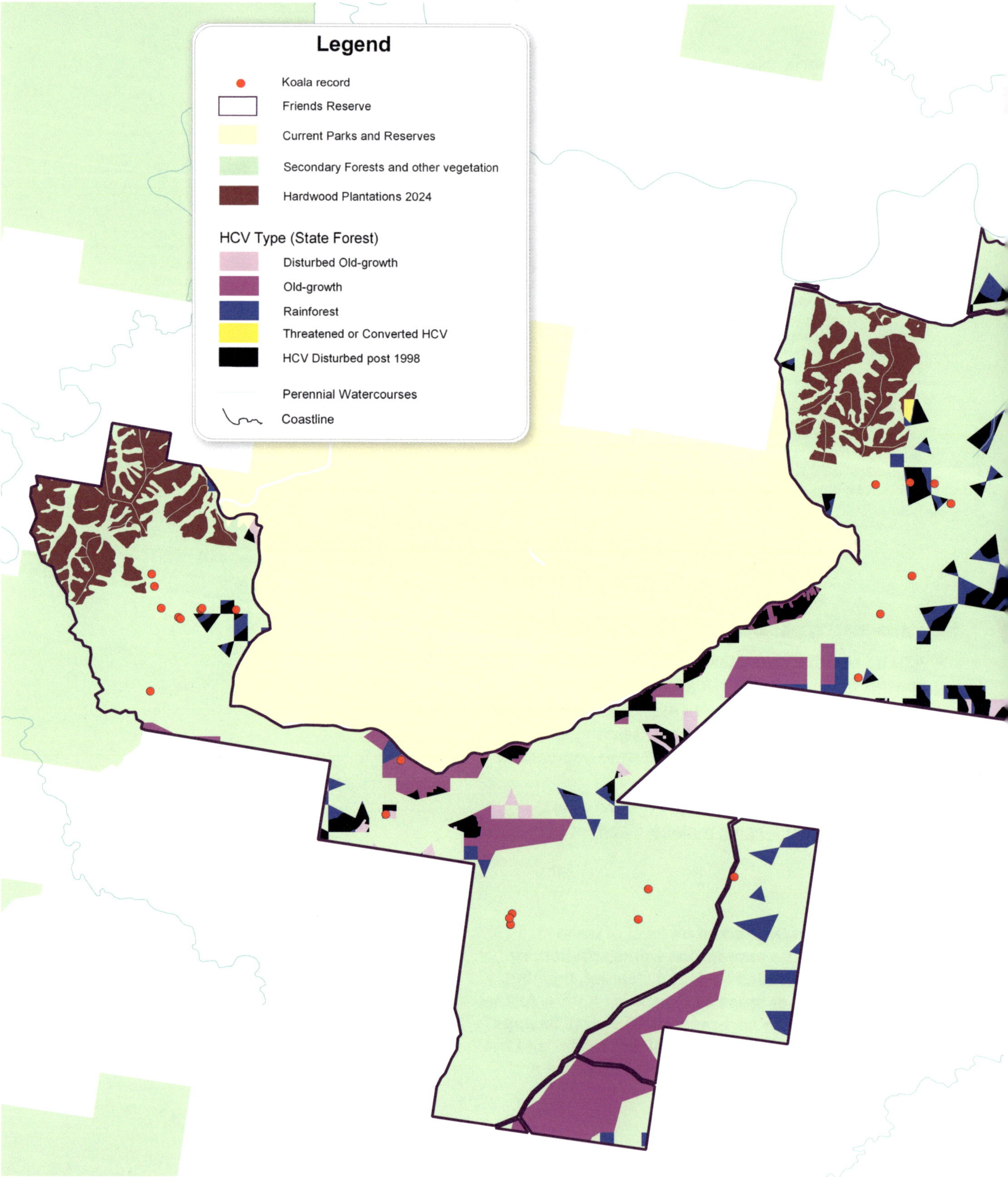

N
0
4 km

FRIENDS OF NEWRY
RESERVE DETAILS

FOREST HABITATS

Newry and Little Newry are a mosaic of many forest types and within them a complex array of vegetation community types owing to the proximity to both the coast and the hinterland, and the variety of elevations, aspects and substrates (basaltic, granitic, sand, and seasonally and permanently inundated alluvium).These include wet forests of the following types:

Wet Sclerophyll - Lowland/Palm, Subtropical, Hinterland, Grassy, and Escarpment.

Rainforest (mostly Subtropical) - Littoral, Lowland, Gallery, and Dry.

Wetlands - Saline, Saltmarsh, Coastal Floodplain, Freshwater, Coastal Freshwater Lagoons, Estuarine, and/or Forested.

Table 9b. FoN - Koala food trees present and regional use rate.

Tallowwood *Eucalyptus microcorys*	High
Grey Gum *Eucalyptus propinqua*	High
Red Mahogany *Eucalyptus resinifera*	High
Blue Gum *Eucalyptus saligna*	Significant
Flooded Gum *Eucalyptus grandis*	Significant
White Mahogany *Eucalyptus acmenoidies*	Significant
Coastal Ironbark *Eucalyptus siderophloia*	Significant
Turpentine *Syncarpia glomulifera*	Significant
Brush Box *Lophostemon confertus*	Significant
Forest Oak *Allocasuarina torulosa*	Significant

* Scat evidence indicates local high preferred use.

THREATENED SPECIES

The forests of Newry and Little Newry host a multitude of threatened species. The varying environment allows for these species to thrive in undisturbed areas. The ancient Rufous Scrub Bird is found at higher altitudes, and Critically Endangered Littoral Rainforest Communities near the coast. Threatened and iconic species such as Koala, Platypus, White-bellied Sea Eagle and Glossy Black-Cockatoo inhabit these forests that are also home to rare and endemic flora species such as the Newry Golden Wattle, the Red Bopple Nut and Native Guava. These species are vulnerable to extinction and also of great interest to domestic and international visitors alike.

REASONS FOR INCLUSION

LANDSCAPE

The reserve represents a large scale critical habitat area for Australia's biodiversity. It is bordered by the Urunga, Valla and Kalang communities and so is adjacent to a large number of private properties. Newry contains the geographical boundary between the Kalang Catchment and the headwaters of Deep Creek. The forest of Newry is critical for the area as its geographical biodiversity supports a range of habitats demonstrated by its coastal associations which graduate into high hinterland vegetation classes. The sheer number of endangered ecological communities and species living within the reserve demonstrates the value of this landscape, and its fragility.

KOALAS

Newry is known to support a significant population of Koalas. A 2019 survey conducted by the University of New South Wales found that Newry's Koala population was one of the largest in the state. What makes Newry forest so successful for Koalas is the presence of many suitable habitat and feed trees dispersed amongst rainforest species (these cool pockets are essential for the Koala's survival in the changing climate). It also has connectivity to five other forests and reserves which have Koala populations present, essential for genetic diversity and population migrations.

CULTURAL HERITAGE

FoN is aware of the impacts of recent forestry operations on Country, and supports the return of culturally appropriate management to the reserve proposal area.

TOURISM AND RECREATION

Today the forest holds many arsenic and gold mine sites from the 1800's, and there are also relics of original bullock and timber equipment. It is shared today with developing respect for First Nations knowledge and management practices, by walking clubs, horseriders, artists, ecologists, professional motocross riders, kids motorbike clubs, photographers, downhill mountain bikers, citizen scientists, and many more local and international visitors alike.

Little Newry

WORLD HERITAGE VALUES

The way people can survive and their interaction with biodiversity in drying climates is important for the world as we enter a new stage of climate change. The continuous population of the ancient forest of Newry holds outstanding universal value for all of humanity. Global resilience comes from protecting places which tell the story of life for the world.

CURRENT ISSUES

Newry is on the point of restoration or simplification. It has features which must not be fragmented or disturbed, such as rainforest areas, hollow bearing and recruitment trees, and threatened ecosystems which also intersperse the forest. It is home to several endemic species which are found only here and which face higher risks of extinction due to their limited distribution. The reserve is situated within the Northern Tablelands Bioregion biodiversity hotspot which are areas known to hold exceptionally high levels of plant and animal diversity, and which are also known to be under significant threat. Rampant development, unsustainable logging and the disturbance of historical mine sites are critical risks to the biodiversity and survival of the reserve.

FUTURE VISION

The establishment of a reserve has the support of local residents, non-governmental organisations and some land management agencies. Community participation in the forest is already working cohesively between local organisations, and the future vision for Newry and Little Newry revolves around sustainability and collaboration between established and new community groups, and First Nations land councils and people. This reserve represents an excellent chance to promote ecotourism due to its accessibility and proximity to residential properties, small towns and beaches which would allow the local community to directly benefit from the natural value of the area.

Orchid, Newry

THREATENED SPECIES

Table 9c. FoN - Threatened species

COMMON NAME	SCIENTIFIC NAME	COMMONWEALTH EPBC ACT STATUS	NSW BC ACT STATUS
MAMMALS			
Koala	*Phascolarctos cinereus*	Comm End	NSW End
Grey-headed Flying Fox	*Pteropus poliocephalus*	Comm Vul	NSW Vul
Yellow-bellied Glider	*Petaurus australis*	Comm Vul	NSW Vul
Squirrel Glider	*Petaurus norfolcensis*		NSW Vul
Greater Glider	*Petauroides volans*	Comm End	NSW End
BIRDS			
Glossy Black-Cockatoo	*Calyptorhynchus lathami lathami*	Comm Vul	NSW Vul
White-bellied Sea Eagle	*Haliaeetus leucogaster*		NSW Vul
Powerful Owl Ninox	*Ninox strenua*		NSW Vul
Masked Owl	*Tyto novaehollandiae*		NSW Vul
Grey-crowned Babbler	*Pomatostomus temporalis temporalis*		NSW Vul
Little Lorikeet	*Glossopsitta pusilla*		NSW Vul
Varied Sitella	*Daphoenositta chrysoptera*		NSW Vul
FLORA			
Newry Golden Wattle	*Acacia chrysostricha*		NSW End
Rusty Plum	*Neimeyera whitei*		NSW Vul
Square-fruited Ironbark	*Eucalyptus tetrapleura*		NSW Vul
Arrowhead Vine	*Tinospora tinosporoides*		NSW End
Native Guava	*Rhodomyrtus psidiodes*	Comm Crit End	NSW Crit End
Scrub Turpentine	*Rhodamnea rubescens*	Comm Crit End	NSW Crit End
Milky Silkpod	*Parsonsia dorrigoensis*	Comm End	NSW Vul
Slender Marsdenia	*Marsdenia longiloba*	Comm End	NSW Vul
Red Bopple Nut	*Hicksbeachia pinnatifolia*	Comm Vul	NSW Vul
Scented Acronychia	*Acronychia littoralis*	Comm End	NSW End
AMPHIBIANS			
Giant Barred Frog	*Myxophyes iteratus*	Comm Vul	NSW Vul
RARE AND ENDEMIC			
Aniseed Myrtle	*Anetholea anisata*		
Nambucca Ironbark	*Eucalyptus fusiformis*		
ENDANGERED ECOLOGICAL COMMUNITIES			
Freshwater Wetlands on Coastal Floodplains of the New South Wales North Coast, Sydney Basin and South East Corner Bioregions.			EEC
Lowland Rainforest in the NSW North Coast and Sydney Basin Bioregions.			EEC
Lowland Rainforest of Subtropical Australia		Comm Crit End	
Lowland Rainforest on Floodplain in the New South Wales North Coast Bioregion.			EEC
Subtropical eucalypt floodplain forest and woodland of the New South Wales North Coast and South East Queensland bioregions.		Comm End	
Swamp Oak Floodplain Forest of the New South Wales North Coast, Sydney Basin and South East Corner Bioregions.			EEC
Swamp Sclerophyll Forest on Coastal Floodplains of the New South Wales North Coast, Sydney Basin and South East Corner Bioregions.			EEC

This globally significant initiative represents an important step toward securing the future of Koalas and the broader ecosystems upon which this endangered species needs to survive and thrive

FRIENDS OF ORARA

VALUES STATEMENT

PHILOSOPHY

The Friends of Orara is a community based group concerned about the deforestation and associated environmental pressures that are increasingly resulting in the destruction and degradation of habitat in the Orara region.

There is a sense of great urgency amongst this community that there will be no coming back from the disconnection of several habitat corridors that have historically provided the needed feed trees and movement opportunities for Koalas and other species.

MISSION

To protect local public forests, habitats and promote sustainable forestry practices.

GOALS

- Preserve habitat and
- Restore green corridors, and thereby
- Ensure the survival of Koalas and many other species in this area.

Citizen science in action

Echidna, Orara East

Red cedar, Boambee

Old-growth habitat tree, Orara East

FRIENDS OF ORARA
RESERVE DETAILS

Table 10a. FoO - Reserve data.

Name of Group:	**FRIENDS OF ORARA**
Total Recorded Koala Sightings	**307**
Total area of reserve (ha)	**10,111**
State Forest	**10,111**
State Forests	**ORARA EAST, ORARA WEST, BOAMBEE**
Area zoned Plantation (ha)	**586**
Area zoned hardwood plantation	**541**
High conservation values (ha)	**3,330**
Rainforest	**1,529**
Old-growth	**1,014**
Disturbed old-growth	**787**
Total disturbed HCV since 1998 (ha)	**2,841**

Indicative assessment area of restoration (ha): 3,400
A large area of erstwhile HCV forest has been damaged since 1998. Recent forestry in the native forest and areas zoned plamtation has done further harm and is a priority for restoration before weed infestations get any worse.

Type of restoration:
Dependent on level of weed infestation and/or site preparation works required, eg machinery may be needed for initial clearing, thinning, ground remediation or preparation works. Annual follow-up maintenance will be necessary (3-10 years). Savings can be made with volunteers contributing labour and by sourcing donations of plants, materials, expertise and grant funding.

FRIENDS OF ORARA

PROPOSED RESERVE

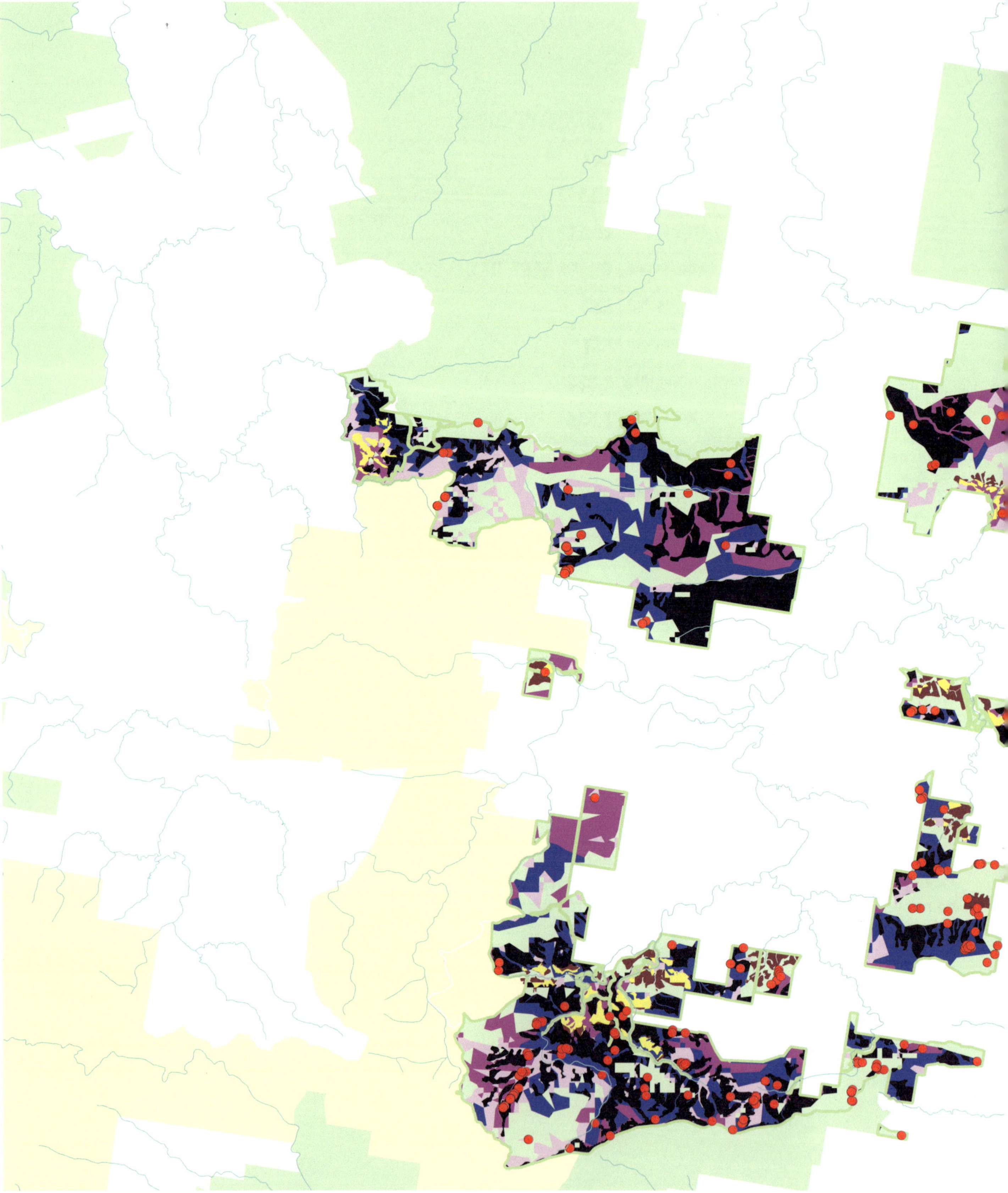

Fig. 9 FoO - Proposed reserve

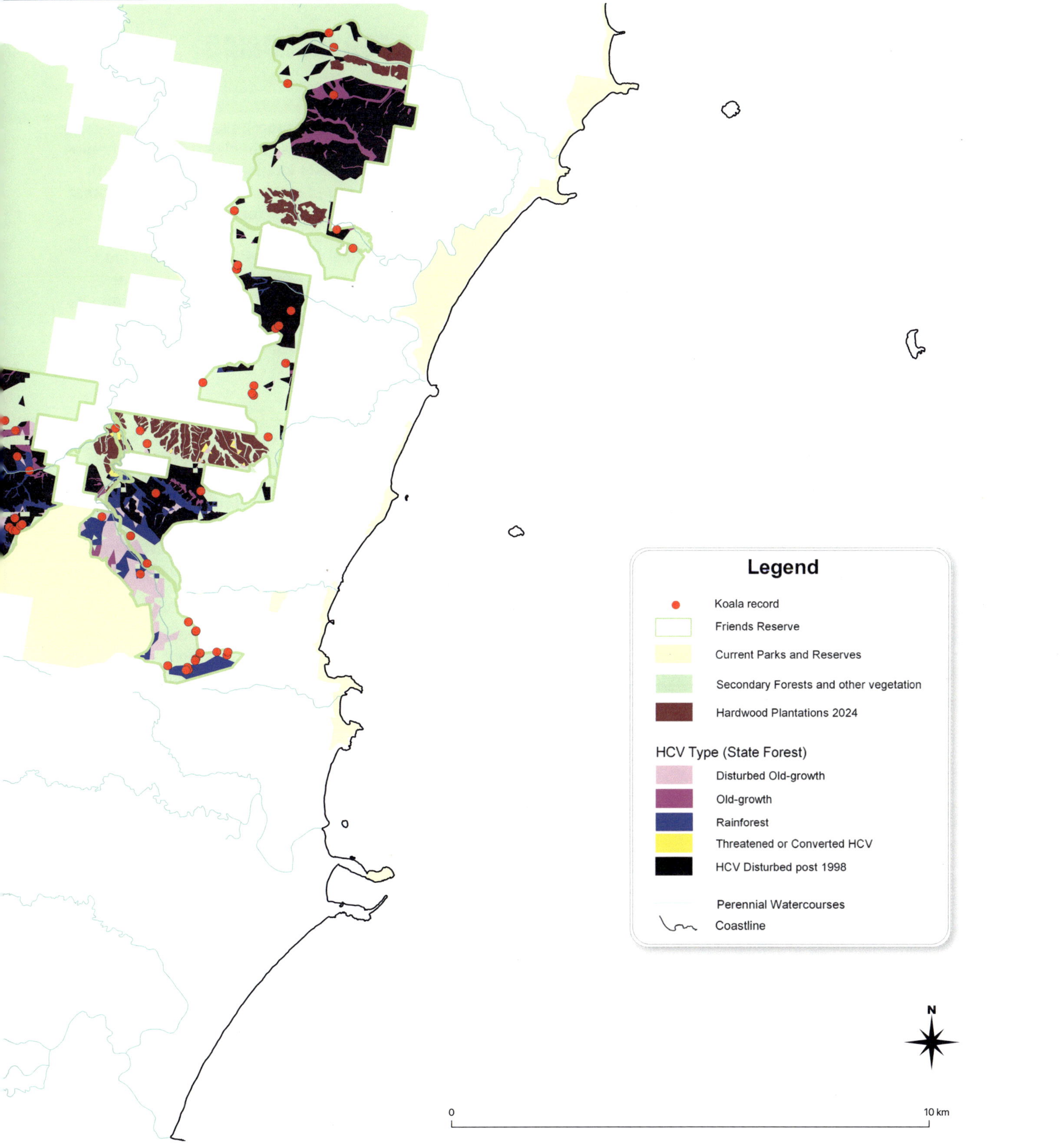

FRIENDS OF ORARA
RESERVE DETAILS

FOREST HABITATS

The forests of the Orara region are predominantly a diverse mix of dry and wet sclerophyll forests with rainforest understory elements in the latter. Wet gullies supporting sub-tropical rainforest communities are a common feature in the region. These forests provide suitable habitat for a diverse range of fauna including a number of species identified as threatened, vulnerable or of special concern.

Species: Tallowwood, Swamp Mahogany, Flooded Gum, Blue Gum, Forest Red Gum, Small-fruited Grey Gum, Pink Bloodwood, Red Mahogany, Forest Oak, Forest Oak, Broad Leaved Paperbark.

Table 10b. FoO - Koala food trees present and regional use rate.

Tallowwood *Eucalyptus microcorys*	High
Grey Gum *Eucalyptus propinqua*	High
Red Mahogany *Eucalyptus resinifera*	High
Blue Gum *Eucalyptus saligna*	Significant
Flooded Gum *Eucalyptus grandis*	Significant
White Mahogany *Eucalyptus acmenoidies*	Significant
Coastal Ironbark *Eucalyptus siderophloia*	Significant
Turpentine *Syncarpia glomulifera*	Significant
Brush Box *Lophostemon confertus*	Significant
Forest Oak *Allocasuarina torulosa*	Significant

THREATENED SPECIES

See Table 10c. FoO - Threatened species

REASONS FOR INCLUSION

LANDSCAPE

The eucalypt forests of the area have been recognised as some of the most diverse and species rich communities in the world. These forests include high levels of both *Eucalyptus* species found nowhere else and the highest diversity of different eucalypts found anywhere.

Endangered Ecological Communities (EEC's) include: Coastal Saltmarsh, Freshwater Wetlands, Littoral Rainforests, Lowland Rainforests, Lowland Rainforests, Subtropical Rainforest Floodplains, White Gum Moist Forests and Swamp Sclerophyll Forest.

KOALAS

The Coffs Harbour—Guy Fawkes meta-population is centered on the Coffs Harbour, northern Bellingen and south-western Clarence Valley LGAs, and extends from the coastal plains at Coffs Harbour/Bongil Bongil National Park west through hinterland and escarpment forests to Guy Fawkes River National Park. This meta-population is considered to be of national significance as a Koala core area.

Connectivity is Key - Orara East, Orara West, and Boambee state forests form part of the Coffs Harbour - Guy Fawkes Koala meta-population. They sit adjacent to the spectacular Bindarri National Park and Ulidarra National Park and Bindarri State Conservation Area. These public forests create an important regional Koala corridor linking remnant and fragmented habitat. Orara West borders prime Koala habitat with Tuckers Nob and Boambee SF providing linkages to one of the coasts richest Koalas populations. The coastal Koala communities of Lower Bucca forests also connect vital habitat corridors from Ulidarra NP and Orara East.

The Great Koala World Heritage Area will protect these unique forests creating a meaningful Koala corridor superhighway, and act as an interchange of dispersing Koalas from two regional populations, the Coffs Harbour and Guy Fawkes meta-population.

CULTURAL HERITAGE

FoO is sensitive to the cultural significance of Country within the reserve. It is also aware of the impacts of development on cultural heritage such as the Pacific

Forest habitat, Orara East

Highway upgrade. FoO therefore supports the protection and culturally appropriate management of all ecosystems, especially the forests and Koala habitat of the region.

TOURISM AND RECREATION

The beginnings of tourism are already well underway in the region. The ecotourism-accredited Bruxner Park Flora Reserve in Orara East is home to the award-winning Forest Sky Pier and the Gumgali Track. Building on these landmarks, walking tracks and mountain-bike paths could be extended into many of the areas outlined in this proposal. The Orara Valley is considered by many to be on-par with the Byron hinterland in terms of natural beauty and tourist potential. Nature tours, short stay accommodation and local produce offerings would flourish if supported in context of a winding down of timber extraction.

WORLD HERITAGE VALUES

Several selection criteria regarding the value of the habitat in this region meet the eligibility for World Heritage assessment. The natural criteria and related values include the requirement for a nominated area to be an outstanding example representing significant on-going ecological and biological processes in the evolution and development of terrestrial, fresh water, coastal and marine ecosystems and communities of plants and animals.

The region represents a unique and complex tapestry of the various stages of ecological transitions across the millenia. Examples of enduring and emerging ecosystems, whilst being threatened, still provide opportunities for preservation and conservation into the future.

To contain the most important and significant natural habitats for in-situ conserva tion of biological diversity, including those containing threatened species of outstanding universal value from the point of view of science or conservation.

The region is a known and well documented stronghold for many endangered species. The pressures of industrial logging and urban development represent the single most urgent issues that need resolution to ensure the survival of not only the iconic Koala but many other species that coexist with this endearing creature.

CURRENT ISSUES

Anthropocene Barriers

Intensified logging and urban development are increasingly impeding Koala movement affecting gene flow and food accessibility. Re-establishment of landscape-level linkages are urgently needed for the survival of this endangered species.

Water security / threatened species

A significant area of High Conservation old-growth forest, with patches of Subtropical Lowland Rainforest are interspersed throughout these forests. Typically for North Coast wet forests in Coffs Harbour, there are numerous streams that feed the mighty Orara River, home to the endangered Eastern Freshwater Cod. Protection of these forests will aid in threatened species survival, ensure future water security and provide refuge and sanctum in climate extremes.

Koala, Orara East

Climate Change / Fire Vulnerability

These forests were not burnt during the fires of Black Spring and Black Summer, as a consequence they are a refuge habitat for many native fauna species badly impacted by the fires, and a place from which populations of these species can rebuild to colonise forests as they recover post-fire.

Bird's-Nest fern, Orara East

FUTURE VISION

Tourism Opportunities

The beginnings of tourism is already underway in the region. The ecotourism-accredited Bruxner Park Flora Reserve is in Orara East State Forest, and is home to the award-winning Forest Sky Pier and the Gumgali Track.

Building on these landmarks, walking tracks and mountain-bike paths could be extended into many of the areas outlined in this proposal.

FoO supports the urgent establishment of the Greater Koala Park World Heritage Area. This globally significant initiative represents an important step toward securing the future of Koalas and the broader ecosystems upon which this endangered species needs to survive and thrive. It aligns with an urgent commitment towards biodiversity preservation, climate change mitigation, and sustainable tourism.

We urge stakeholders to support this proposal, recognising the ecological, cultural, and economic value it brings to the region. Together, we can ensure a brighter future for Koalas living wild and healthy, protected in the towering eucalypt forests of NSW Mid North Coast for all the world to treasure.

Transitioning the Timber Industry

Any loss in employment will be greatly reduced under a program of re-employment in plantation establishment on previously-cleared agricultural freehold, and new jobs in forest restoration and land management.

THREATENED SPECIES

Table 10c. FoO - Threatened species

COMMON NAME	SCIENTIFIC NAME	COMMONWEALTH EPBC ACT STATUS	NSW BC ACT STATUS
FAUNA			
MAMMALS			
Koala	*Phascolarctos cinereus*	Comm End	NSW End
Southern Greater Glider	*Petauroides volans*	Comm End	NSW End
Yellow-bellied Glider	*Petaurus australis*	Comm Vul	NSW Vul
Golden-tipped Bat	*Phoniscus papuensis*		NSW Vul
Parma Wallaby	*Notamacropus parma*		NSW Vul
Spotted-tailed Quoll	*Dasyurus maculatus*	Comm End	NSW Vul
Southern Greater Glider	**Macropus dorsalis*		NSW End
Brush-tailed Rock-wallaby	*Petauroides volans*	Comm Vul	NSW End
Brush-tailed Phascogale	**Petrogale penicillata*		NSW Vul
Red-legged Pademelon	*Phascogale tapoatafa*		NSW Vul
Spotted-tailed Quoll	*Dasyurus maculatus*	Comm End	NSW Vul
Grey-headed Flying Fox	*Pteropus poliocephalus*	Comm Vul	NSW Vul
Eastern False Pipistrelle	*Falsistrellus tasmaniensis*		NSW Vul
Eastern Coastal Free-tailed Bat	*Micronomus norfolkensis*		NSW Vul
Large Bent-winged Bat	*Miniopterus orianae*		NSW Vul
Southern Myotis	*Myotis macropus*		NSW Vul
AMPHIBIANS			
Pouched Frog	*Assa darlingtonii*	Comm Vul	NSW Vul
Peppered Tree Frog	**Litoria piperata*	Comm Vul	NSW End
Stuttering Frog	*Mixophyes balbus*	Comm Vul	NSW End
Giant Barred Frog	*Mixophyes iteratus*	Comm Vul	NSW Vul
Sphagnum Frog	*Philoria sphagnicolus*	Comm Vul	NSW Vul
REPTILES			
Broad-headed Snake	**Hoplocephalus bungaroides*	Comm End	NSW End
Stephens' banded Snake	*Hoplocephalus stephensii*		NSW Vul
BIRDS			
Regent Honeyeater (found further south)	*Anthochaera phrygia*	Comm Crit End	NSW Crit End
Glossy Black-Cockatoo	*Calyptorhynchus lathami lathami*		NSW Vul
Red Goshawk	*Erythrotriorchis radiatus*	Comm End	NSW End
White-bellied Sea Eagle	*Haliaeetus leucogaster*		NSW Vul
Little Eagle	*Hieraaetus morphnoides*		NSW Vul
Eastern Osprey	*Pandion cristatus*		NSW Vul
Black-necked Stork	*Ephippiorhynchus asiaticus*		NSW End

Little Lorikeet	*Glossopsitta pusilla*		NSW Vul
White-throated Needletail	*Hirundapus caudacutus*	Comm Vul	NSW Vul
Cotton Pygmy-Goose	**Nettapus coromandelianus*	Protected	
Australian Painted Snipe	*Rostratula australis*	Comm End	NSW End
Coastal Emu (Evans Head-Red Rock-Bungwalbin)	**Dromaius novaehollandiae*		NSW End
Barking Owl	*Ninox connivens*		NSW Vul
Powerful Owl	*Ninox strenua*		NSW Vul
Masked Owl	*Tyto novaehollandiae*		NSW Vul
Sooty Owl	*Tyto tenebricosa*		NSW Vul
Brown Treecreeper (eastern subspp)	*Climacteris picumnus victoriae*	Comm Vul	NSW Vul
Barred Cuckoo Shrike	*Coracina lineata*		NSW Vul
Wompoo Fruit Dove	*Ptilinopus magnificus*		NSW Vul
Rose-crowned Fruit Dove	*Ptilinopus regina*		NSW Vul
Square-tailed Kite	*Lophoictinia isura*	Comm Rare	NSW Vul
INSECTS			
Giant Dragonfly	**Petalura gigantea*		NSW End
Coastal Petaltail	*Petalura litorea*		NSW End
Southern Pink Underwing Moth	*Phyllodes imperialis*	Comm End	NSW End
FLORA			
Rusty Plum	*Niemeyera whitei*		NSW Vul
Orara Boronia	*Boronia umbellata*	Comm Vul	NSW Vul
Scrub Turpentine	*Rhodamnia rubescens*	Comm Crit End	NSW Crit End
Slender Marsdenia	*Marsdenia longiloba*	Comm Vul	NSW End
Spider Orchid	*Dendrobium melaleucaphilum*		NSW End
Square-stemmed Spike-rush	*Eleocharis tetraquetra*		NSW End
Slender Screw Fern	*Lindsaea incisa*		NSW End
Milky Silkpod	*Parsonsia dorrigoensis*	Comm End	NSW Vul
Moonee Quassia	*Quassia spp. Moonee Creek*	Comm End	NSW End
Ravine Orchid	*Sarcochilus fitzgeraldii*	Comm Vul	NSW Vul
Hartmann's Sarcochilus	*Sarcochilus hartmannii*	Comm Vul	NSW Vul
Austral Toadflax	*Thesium australe*	Comm Vul	NSW Vul
Cryptic Forest Twiner	*Tylophora woollsii*	Comm End	NSW End
Stink Lily	*Typhonium sp. aff.brownii*		NSW End
Native Guava	*Rhodomyrtus psidioides*	Comm Crit End	NSW Crit End
Red Boppelnut	*Hicksbeachia pinnatifolia*	Comm Vul	NSW Vul

ENDANGERED ECOLOGICAL COMMUNITIES

Lowland Rainforest in the NSW North Coast and Sydney Basin Bioregions	Comm Crit End	
Subtropical Coastal Floodplain Forest of the New South Wales North Coast Bioregion		EEC

Save the Tarkeeth Koalas from extinction... through creation of the Greater Koala Park

FRIENDS OF
TARKEETH KOALAS

VALUES STATEMENT

PHILOSOPHY

Friends of Tarkeeth Koalas is a community based alliance of forest friends and neighbours responding to the need to protect biodiversity values and save the endangered Koalas of Tarkeeth.

Our alliance draws upon the earlier work of the Tarkeeth Forest community. We express our gratitude to the 16,000 people who supported our petitions to the NSW government calling for an end to unsustainable forestry operations at Tarkeeth.

MISSION

Our mission is to protect, promote and preserve the natural and cultural heritage of Tarkeeth and to save the Tarkeeth Koalas
from extinction.

GOALS

Achieve permanent protection of Tarkeeth through creation of the Greater Koala Park World Heritage Area.

Frog, Tarkeeth forest

Wompoo Fruit Dove, Tarkeeth forest

Koala, Tarkeeth forest

FRIENDS OF TARKEETH KOALAS
RESERVE DETAILS

Table 11a. FoTK - Reserve data.

Name of Group:	**FRIENDS OF TARKEETH KOALAS**
Total Recorded Koala Sightings	**19**
Total area of reserve (ha)	**1,418**
State Forest	**1,418**
State Forest	**TARKEETH**
Area zoned Plantation (ha)	**854**
Area zoned hardwood plantation	**566**
High conservation values (ha)	**140**
Rainforest	**50**
Old-growth	**87**
Disturbed old-growth	**3**
Total disturbed HCV since 1998 (ha)	**33**

Indicative assessment area of restoration (ha): 1,000

- Plantation: Full or partial restoration. (Species: Blackbutt, Blue Gum, Flooded Gum, Tallowwood. Koala food trees to be retained.)
- CIFOA: Regeneration.
- Other: Assisted regeneration as required.

Type of restoration:

Dependent on level of weed infestation and/or site preparation works required, eg machinery may be needed for initial clearing, thinning, ground remediation or preparation works. Annual follow-up maintenance will be necessary (3-10 years). Savings can be made with volunteers contributing labour and by sourcing donations of plants, materials, expertise and grant funding.

FRIENDS OF TARKEETH KOALAS

PROPOSED RESERVE

Fig. 10 FoTK - Proposed reserve

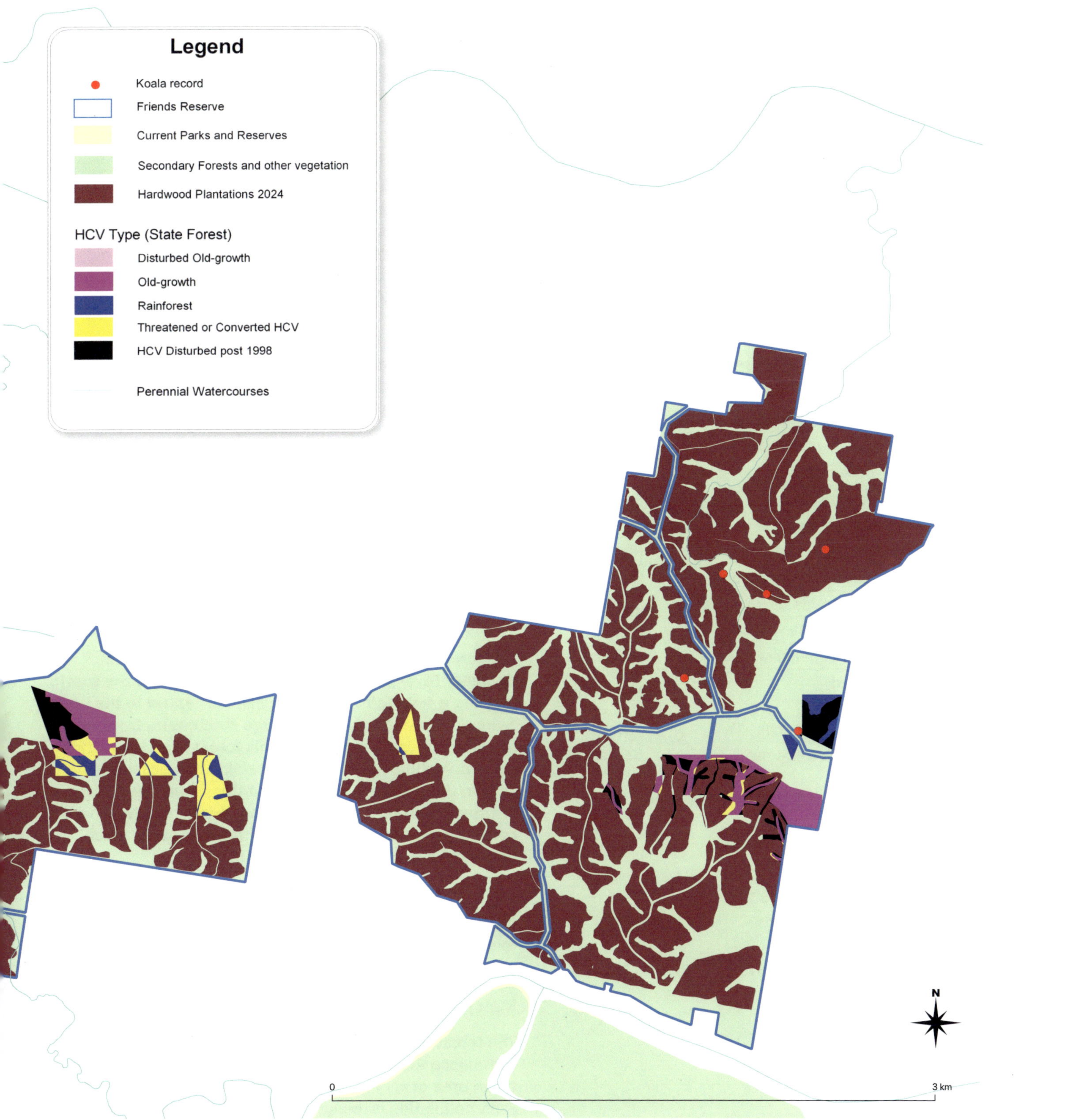

FRIENDS OF TARKEETH KOALAS
RESERVE DETAILS

FOREST HABITATS

The reserve contains a matrix of North Coast and Northern Hinterland Wet Sclerophyll eucalypt dominated forests, including Northern Lowland Subtropical Rainforest and pockets of high conservation value, old-growth and swamp interspersed. The watershed and wetland systems of the reserve are interconnected with a range of forest habitats and the Bellinger- and Kalang River catchments.

Threatened and Endangered Ecological Systems: Northern Lowland Subtropical Rainforest (TEC/EEC), Swamp Sclerophyll Forest on Coastal Floodplains (TEC). Koala food trees:

Table 11b. FoTK - Koala food trees present and regional use rate.

Tallowwood *Eucalyptus microcorys**	High
Small-fruited Grey Gum *Eucalyptus propinqua**	High
Forest Oak *Allocasuarina torulosa**	Significant
Blue Gum *Eucalyptus saligna*	Significant
Flooded Gum *Eucalyptus grandis*	Significant
Blackbutt *Eucalyptus pilularis*	Significant
Red Mahogany *Eucalyptus resinifera*	High
White Mahogany *Eucalyptus acmenoidies*	Significant
Coastal Ironbark *Eucalyptus siderophloia*	Significant
Swamp Mahogany *Eucalyptus robusta*	High
Turpentine *Syncarpia glomulifera*	Significant
Broad-leaved Paperbark *Melaleucia quinquenervia*	Significant
Pink Bloodwood *Corymbia intermedia*	Occasional
Brush Box *Lophostemon confertus*	Occasional

* Scat evidence indicates local high preferred use.

THREATENED SPECIES

Tarkeeth supports a wide variety of threatened and endangered species.

Table 11c. FoTK - Threatened species

REASONS FOR INCLUSION

Tarkeeth lies on Gumbaynggirr Country at the heart of the proposed Greater Koala Park World Heritage Area, on the iconic rim of the Bellinger Valley at the easterly extension of the Great Dividing Range.

This forest should be protected as a reserve due to its unique position above and separating the Bellinger and Kalang Rivers, and encompassing a large area within the forested landscape of the Urunga Escarpment Moist Climate Change Corridor, identified as key habitat required for species to adapt to climate change.

This reserve would provide a vital link in the landscape for threatened species, notably the Koala, between the Urunga coastal lowlands, World Heritage-listed Gondwana Rainforests of the Dorrigo escarpment, and the larger habitats of the western hinterlands.

Historical and contemporary records confirm the continued presence of Koalas across this landscape. This Koala metapopulation is endangered and recognised as internationally significant.

LANDSCAPE

The forest rises on Fernmount Range separating the Bellinger River to the north and Kalang River to the south, forming a forested watershed which is central to the upper Bellinger-Kalang estuary drainage catchment. Owing to its location above the two rivers, this forest is an important rain-harvesting, filtration, and stabilization ecosystem vital to the wetlands and rivers of this estuarine system. Nearby, the lower estuary is formed where the two rivers converge, creating a large and dynamic system which flows into the Pacific Ocean at Urunga.

A unique feature of this landscape is its forested ridge between the two rivers linking the Bellingen and Urunga townships and Bellinger-Kalang Estuary environs.

KOALAS

The forest is encompassed within the North Macleay-Nambucca Area of Regional Koala Significance, identified as an area of very high Koala occupancy. This Koala metapopulation is recognised as internationally significant.

Glossy Black-Cockatoo

Tarkeeth forest lies within a continuous forested landscape. The adjoining private native forests and Tarkeeth Forest Wildlife Refuge (NSW BCT) are identified as 'Preferred Koala Habitat' within the Bellingen Shire Coastal Area Koala Planning Area.

Koala food trees are present throughout the forest matrix. A diversity of landscapes, including pockets of rainforest vegetation and larger old growth trees, provide refuges for Koalas to survive heatwaves.

Forestry plans and community sightings confirm a resident population of Koalas. Numerous recent records within and around the forest, including sightings of healthy breeding females with young, attest to population recovery, growth and the suitability of the landscape for Koalas.

The relatively fewer number of historical on-site records reflects the absence of targeted Koala surveys and under-reporting due to poor access.

CULTURAL HERITAGE

Tarkeeth has a multitude of both tangible (archaeological) and intangible (cultural) sites and features across the landscape.

TOURISM AND RECREATION

Located at the heart of the proposed Greater Koala Park World Heritage Area, 25km east of the World Heritage-listed Gondwana Rainforests of Dorrigo National Park and just a short stroll from the heritage-listed buildings of the picturesque valley town of Bellingen, the reserve offers extraordinary potential for recreation, nature-based activities, and opportunities for immersion in high cultural awareness and learning.

The forested ridge, which links the Bellingen and Urunga townships and Bellinger-Kalang Estuary environs. In consultation, walking and riding tracks could be established quickly and at relatively low cost using existing trails and access points. A cycleway could be added to link the eastern trail with the Urunga township and the seaside.

The reserve has several potential sites for scenic lookouts offering spectacular views of the Great Escarpment of the Dorrigo Plateau, and the Pacific Coast.

Sites of European Heritage significance exist within and adjoining the reserve, including the Tarkeeth Quarry established in 1892 to supply stone for construction of the Urunga breakwater, and Twin Pines, an early European settlement, the latter being an ideal location for picnic areas, carparking, and recreational access to the Kalang River, with potential for a river crossing service to Newry and Urunga.

FRIENDS OF TARKEETH KOALAS
RESERVE DETAILS

WORLD HERITAGE VALUES

Tarkeeth should become a World Heritage Area due to its unique and outstanding value as a landform connecting The Great Dividing Range and Pacific Ocean, and as a complex watershed system encompassing different topographies and separating two major rivers, which converge and flow into the ocean together.

Tarkeeth is a biodiverse Australian ecosystem that provides critical habitat to numerous endangered species, notably the Koala, and forms a vital link in the landscape between the eastern coastal lowlands, World Heritage-listed Gondwana Rainforests, and larger habitats of the western hinterlands.

Tarkeeth represents a landscape of high cultural significance. FoTK supports the assessment, protection and preservation of this very special place.

CURRENT ISSUES

A major current issue is logging, particularly extreme extractive forestry methodologies. Clearfelling, biomass extraction, logging fires, herbicide poisoning, and replanting with monocultures disrupt, displace and kill native species leaving 'broken links' in the landscape. Urgent restoration of these areas is needed for the survival of endangered species, notably the Koala.

Management of the forest solely as a timber production resource is detrimental to other important values (ecological, cultural, social, visual); suppresses eco-tourism and economic potential; and changes the landscape in a way that increases the risk of high severity bushfire.

FUTURE VISION

Located only a kilometre from Bellingen township and linking with Urunga and the Bellinger-Kalang estuary environs, the benefits of Tarkeeth becoming a protected reserve with recreational facilities and opportunities for nature-based activities and employment will be transformational for surrounding residents, local communities and First Nations people.

Protective and collaborative restoration and management of Tarkeeth as a reserve will ensure this biodiverse Australian ecosystem continues to provide critical habitat to numerous endangered species, notably the Koala.

Koala, Tarkeeth

Tarkeeth represents an outstanding opportunity for economic development through sustainable ecotourism due to its close proximity to Bellingen and Urunga townships, and short driving distance from the World Heritage-listed Gondwana Rainforests of Dorrigo National Park, beautiful beaches of the Coffs Coast and Coffs Harbour Airport.

Friends of Tarkeeth Koalas support the urgent establishment of the Greater Koala Park World Heritage Area.

THREATENED SPECIES

Table 11c. FoTK - Threatened species

COMMON NAME	SCIENTIFIC NAME	COMMONWEALTH EPBC ACT STATUS	NSW BC ACT STATUS
MAMMALS			
Brush-tailed Phascogale	*Phascogale tapoatafa*		NSW Vul
Golden-tipped Bat	*Kerivoula papuensis*		NSW Vul
Grey headed Flying Fox	*Pteropus poliocephalus*	Comm Vul	NSW Vul
Southern Myotis	*Myotis macropus*		NSW Vul
Little Bent-winged Bat	*Miniopterus australis*		NSW Vul
Koala	*Phascolarctos cinereus*	Comm End	NSW End
Southern Greater Glider	*Petauroides volans*	Comm End	NSW End
Spotted-tailed Quoll	*Dasyurus maculatus*	Comm End	NSW Vul
AMPHIBIANS			
Giant Barred Frog	*Mixophyes iteratus*	Comm End	NSW End
BIRDS			
Glossy Black-Cockatoo	*Calyptorhynchus lathami*	Comm Vul	NSW Vul
Little lorikeet	*Glossopsitta pusilla*		NSW Vul
Masked Owl	*Tyto novaehollandiae*		NSW Vul
Powerful Owl	*Ninox strenua*		NSW Vul
Rose-crowned Fruit-dove	*Ptilinopus regina*		NSW Vul
Sooty Owl	*Tyto tenebricosa*		NSW Vul
Square-tailed Kite	*Lophoictinia isura*		NSW Vul
Superb Fruit-dove	*Ptilinopus superbus*		NSW Vul
White-bellied Sea-Eagle	*Haliaeetus leucogaster*		NSW Vul
Wompoo Fruit Dove	*Ptilinopus magnificus*		NSW Vul
FLORA			
Milky Silkpod	*Parsonsia dorrigoensis*	Comm End	NSW Vul
Native Guava	*Rhodomyrtus psidioides*	Comm Crit End	NSW Crit End
Red Bopple Nut	*Hicksbeachia pinnatifolia*		NSW Vul
Rusty Plum	*Niemeyera whitei*		NSW Vul
Scrub Turpentine	*Rhodamnia rubescens*	Comm Crit End	NSW Crit End
Slender Marsdenia	*Marsdenia longiloba*	Comm Vul	NSW End
ENDANGERED ECOLOGICAL COMMUNITIES			
Lowland Rainforest of Subtropical Australia		Comm Crit End	
Coastal Swamp Sclerophyll Forest of New South Wales and South East Queensland		Comm End	
Swamp Sclerophyll Forest on Coastal Floodplains of the New South Wales North Coast, Sydney Basin and South East Corner Bioregions			EEC
Lowland Rainforest in the NSW North Coast and Sydney Basin Bioregions			EEC

To have Tuckers Nob protected permanently as vital Koala habitat and a critical central forest within the Greater Koala Park

FRIENDS OF TUCKERS NOB

VALUES STATEMENT

Moist forest, Tuckers Nob

Aniseed Myrtle (Ringwood), Tuckers Nob

PHILOSOPHY

Friends of Tuckers Nob is a network of community members working to promote and ensure the protection of the forests and Koalas of Tuckers Nob. Friends of Tuckers Nob (FoTN) collaborate on information exchange, surveying, advocacy and community events. We prioritize an inclusive and apolitical practice of working together.

MISSION

FoTN aim to have Tuckers Nob protected permanently as vital Koala habitat and a critical central forest within the Greater Koala Park World Heritage Area.

We aim to assist in the mitigation of climate change, reduce local fire hazard, and preserve the water quality and security for Never Never River.

GOALS

Survey, record and map the forests and wildlife of Tuckers Nob.

Promote an awareness and understanding of the significance of Tuckers Nob forest.

Organize events and art projects with the intention of building a sense of identity and connection to the Koala as an animal of importance for the local community.

Ringwood (Aniseed Myrtle), Tuckers Nob

FRIENDS OF TUCKERS NOB
RESERVE DETAILS

Table 12a. FoTN - Reserve data.

Name of Group:	**FRIENDS OF TUCKERS NOB**
Total Recorded Koala Sightings	**75**
Total area of reserve (ha)	**4,375**
State Forest	**4,375**
State Forest	**TUCKERS NOB**
Area zoned Plantation (ha)	**2,457**
Area zoned hardwood plantation	**1,678**
High conservation values (ha)	**505**
Rainforest	**206**
Old-growth	**170**
Disturbed old-growth	**129**
Total disturbed HCV since 1998 (ha)	**152**

Indicative assessment area of restoration (ha): 1,500
Almost 1,000 ha of healthy Koala habitat zoned plantation have been cleared in in recent years. Areas of rainforest and other forest converted during initial plantation establishment and subsequently, need to be restored.

Type of Regeneration:
Dependent on the level of weed infestation and/or site preparation works required, eg. if machinery is needed for initial clearing of weeds or for other preparatory ground works. A minimum 3 years follow-up maintenance work is necessary, ideally continuing for 10 years. Savings can be made with volunteers contributing labour, materials and expertise, and by sourcing donations and grant funding.

FRIENDS OF TUCKERS NOB

PROPOSED RESERVE

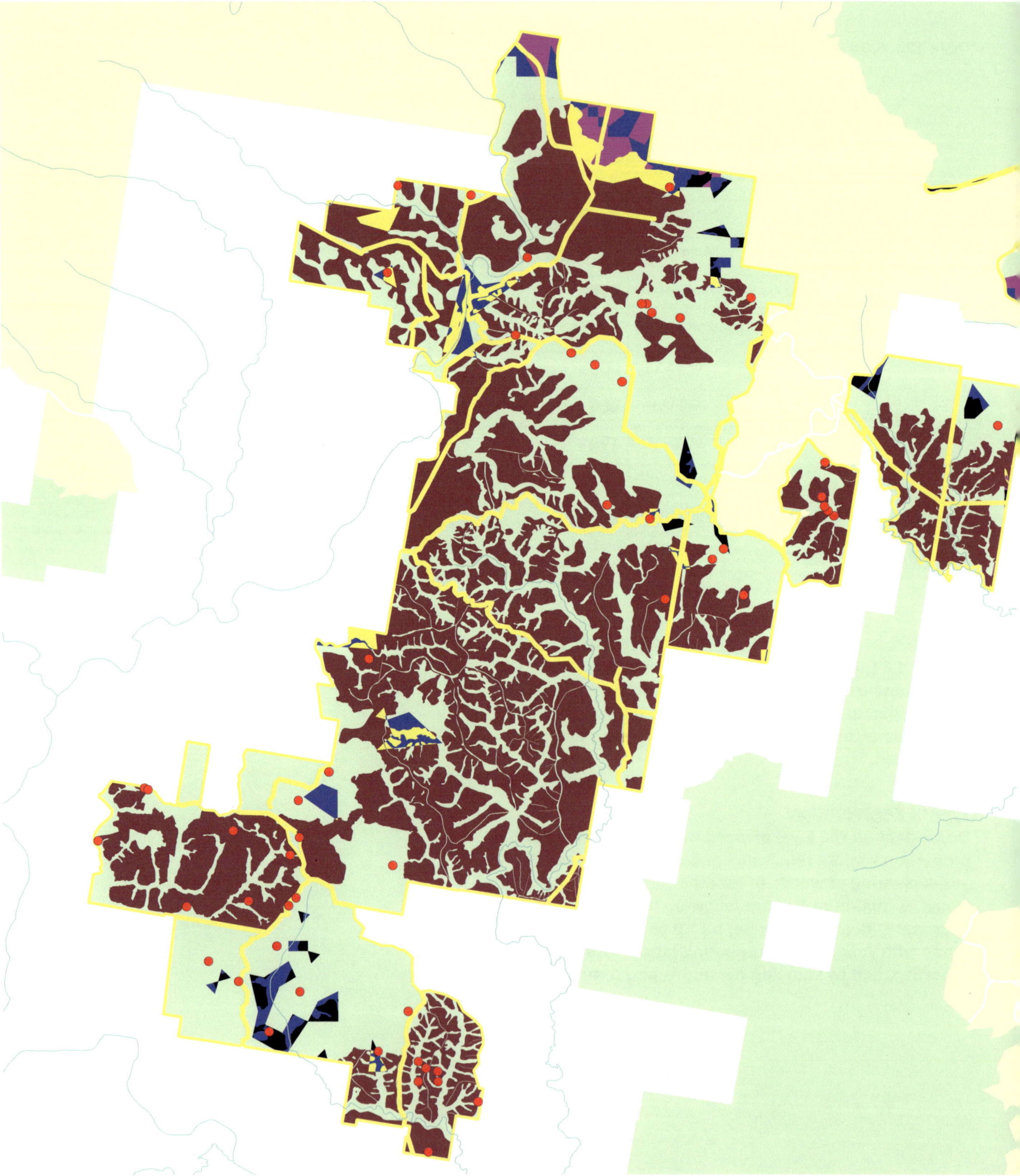

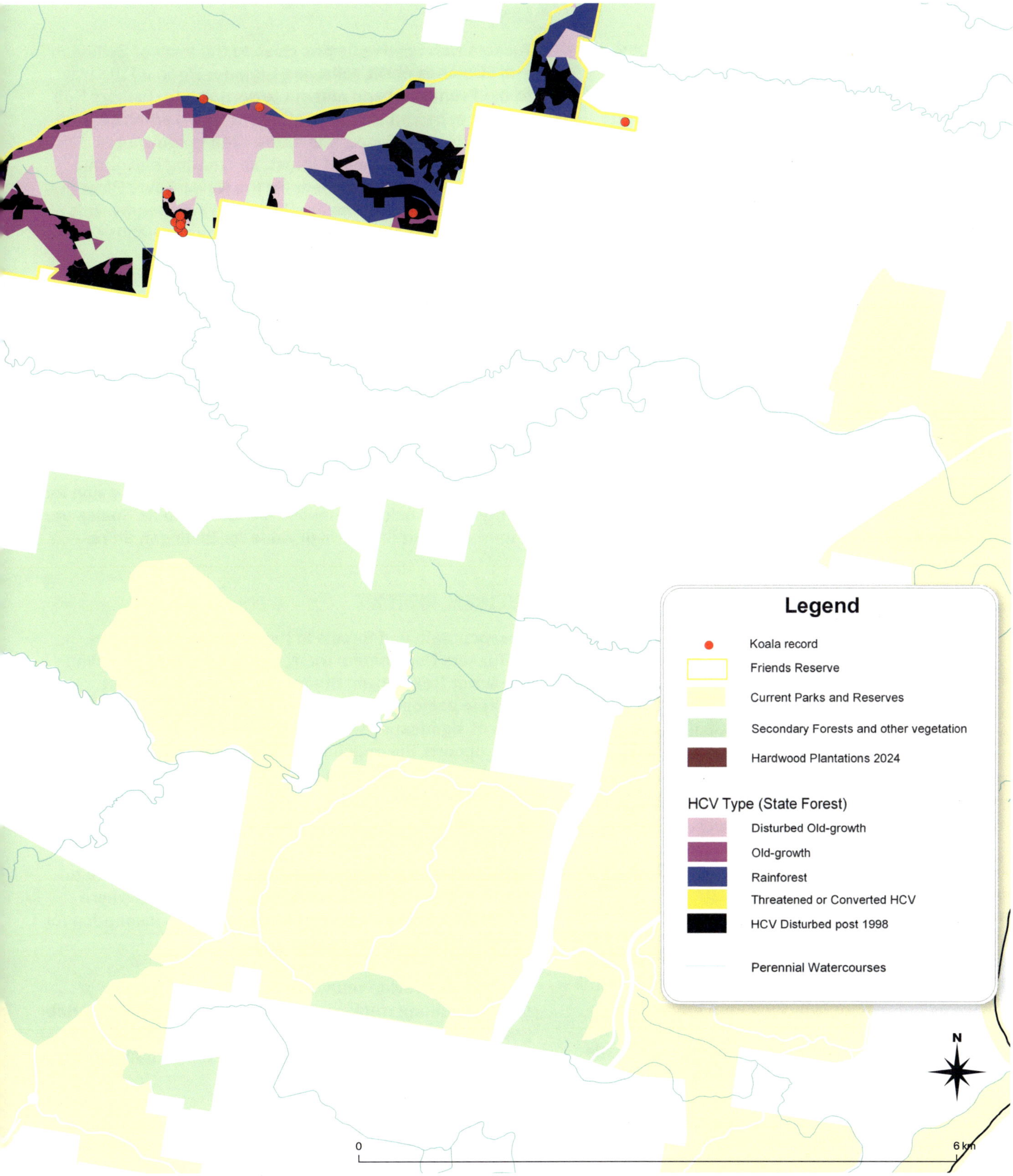
Legend
Koala record
Friends Reserve
Current Parks and Reserves
Secondary Forests and other vegetation
Hardwood Plantations 2024
HCV Type (State Forest)
Disturbed Old-growth
Old-growth
Rainforest
Threatened or Converted HCV
HCV Disturbed post 1998
Perennial Watercourses
N
0
6 km

FRIENDS OF TUCKERS NOB
RESERVE DETAILS

FOREST HABITATS

Tuckers Nob is an area of exceptional diversity with a wide range of ecological communities and habitats. Established eucalypt forests, including areas of rainforest and old-growth stretch across the reserve.

Table 12b. FoTN - Koala food trees present and regional use rate.

Tallowwood *Eucalyptus microcorys*	High
Grey Gum *Eucalyptus propinqua*	High
Red Mahogany *Eucalyptus resinifera*	High
Blue Gum *Eucalyptus saligna*	Significant
Flooded Gum *Eucalyptus grandis*	Significant
White Mahogany *Eucalyptus acmenoidies*	Significant
Coastal Ironbark *Eucalyptus siderophloia*	Significant
Turpentine *Syncarpia glomulifera*	Significant
Brush Box *Lophostemon confertus*	Significant
Forest Oak *Allocasuarina torulosa*	Significant

THREATENED SPECIES

Tuckers Nob supports many species recognised as threatened.

See table 12c. FoTN - Threatened species

REASONS FOR INCLUSION

Tuckers Nob is located in the centre of the proposed Greater Koala Park World Heritage Area. It is a critical linking forest for Koalas, providing connection between the hinterland and plateau habitats of the World Heritage listed Dorrigo National Park, Bindarri NP and Orara, to the coast with Pine Creek and Bongil Bongil NP.

Due to its geographical position on the NSW North Coast, a bioregion that lies in a transitional zone changing from coastal to the Great Dividing Range, the forest has a high diversity of species. Tuckers Nob reserve features healthy eucalypt forest, remnant and emerging rainforest. Areas of old-growth forest provide essential habitat for a number of species dependant on hollow-bearing trees.

The area has a notable number of mature Aniseed Myrtle or Ringwood trees, a species found only in the Nambucca, Kalang and Bellinger valleys. This species is now under threat from the introduced myrtle rust, and has been submitted to the New South Wales Threatened Species Scientific Committee to be considered for listing as endangered.

LANDSCAPE

Tuckers Nob reserve begins close to the town of Bellingen and stretches north, following Gleniffer Road all the way to the Promised Land and out toward Valery. Tuckers Nob mountain has an elevation of 920m with a small summit area and steep slopes. The entire area is convuluted with ridges and gullies of forest, all creating a dynamic environment and catchment for the Never Never River that winds through and joins the Bellinger River nearby. This environment is classified as Class 1 Aquatic Habitat.

KOALAS

Tuckers Nob is recognized as core Koala habitat supporting a healthy population of Koalas. The Koalas of Tuckers Nob are vital for the survival of the local sub-populations. Without them, Bindarri NP would be deprived of a key source population and could dissappear within a matter of years due to in-breeding and disease.

Tuckers Nob is a central part of an interconected vision for corridors and links at a landscape level, and its Koalas are a tourism asset of very high value for Bellingen Shire.

CULTURAL HERITAGE

The mountain and forests in the environs now known as Tuckers Nob contain many natural features including the Never Never River, Christmas Holes and ancient volcanic geology. FoTN notes the identification of several sites of signifcance within the proposal. It acknowledges and supports the ongoing connection to, and use of this Country by the Gumbaynggirr peoples.

TOURISM AND RECREATION

Tuckers Nob has many recreational values appealling to tourism. Situated close to the town of Bellingen where music and arts are a focus, the forest offers inspiration for writers, musicians and visual arts.

The forests provide many opportunities for a range of diverse walking tracks and swimming spots. Tuckers Nob peak is known and favoured as a running track. Nearby is the start of the Syndicate Ridge track, a challenging walk to the Dorrigo Plateau.

Supporting the protection of the local Koala population, and the potential of spotting them in the wild is a strong tourism drawcard.

Mature forest, Tuckers Nob

WORLD HERITAGE VALUES

Tuckers Nob is both connected to, and an outlier of, the Great Escarpment, and a part of the Ebor Volcano caldera.

It has been mapped as a coastal corridor in the context of climate change, providing safe movement of fauna to refuges in cooler altitudes and microclimates.

Healthy Koala habitat and healthy Koalas are a feature of Tuckers Nob forest. The inclusion of these forests within the Greater Koala Park World Heritage Area will create a secure passage for Koalas, providing a vital link in a continuous forested corridor from the coast across the Great Eastern Range, to the tablelands. Inclusion of the large connected forest component to the south of the proposed reserve is critical to the survival of Koala populations across the Bellinger valley and beyond.

CURRENT ISSUES

The active encroachment of forestry activities into Tuckers Nob is causing great stress to the local community due to loss of amenity, increased fire risk and impacts on water quality. Mature and old-growth trees have been found throughout the areas currently being clear-felled and proposed for clear-felling. Unique and special biodiversity values beyond the Koala exisit within the forest, including original remnant forest and endemic rainforest species such as the Aniseed Myrtle, which must be protected.

FUTURE VISION

The forests and peak of Tuckers Nob, Christmas Holes, Never Never River and the Promised Land Loop are internationally signifciant and would attract a steady industry of sustainable eco-tourism with a wild Koala population as a key asset.

RESERVE DETAILS

THREATENED SPECIES

Table 12c. FoTN - Threatened species

COMMON NAME	SCIENTIFIC NAME	COMMONWEALTH EPBC ACT STATUS	NSW BC ACT STATUS
FAUNA			
MAMMALS			
Koala	Phascolarctos cinereus	Comm End	NSW Vul
Spotted-tailed Quoll	Dasyurus maculatus		NSW Vul
Brush-tailed Phascogale	Phascogale tapoatafa		NSW Vul
Golden-tipped Bat	Phoniscus papuensis		NSW Vul
Little Bent-wing Bat	Miniopterus australis		NSW Vul
Large Bent-wing Bat	Miniopterus oranae		NSW Vul
Grey-headed Flying Fox	Pteropus poliocephalus	Comm Vul	NSW Vul
Southern Greater Glider	Petauroides volans	Comm End	NSW End
Yellow-bellied Glider	Petaurus australis		NSW Vul
BIRDS			
Glossy Black-Cockatoo	Calyptorhynchus lathami		NSW Vul
Red Goshawk	Erythrotriorchis radiatus	Comm End	NSW Crit End
Square-tailed Kite	Lophoictinia isura		NSW Vul
Black-necked Stork	Ephippiorhynchus asiaticus		NSW End
Barking Owl	Ninox connivens		NSW Vul
Powerful Owl	Ninox strenua		NSW Vul
Masked Owl	Tyto novaehollandiae		NSW Vul
Sooty Owl	Tyto tenebricosa		NSW Vul
Regent Honeyeater	Anthochaera phrygia	Comm Crit End	NSW Crit End
Eastern Bristlebird	Dasyornis brachypterus	Comm End	NSW End
White-throated Needletail	Hirundapus caudacutus	Comm Vul	NSW Vul
Wompoo Fruit-dove	Ptilinopus magnificus		NSW Vul
Rose-crowned Fruit-dove	Ptilinopus regina		NSW Vul
Superb Fruit-dove	Ptilinopus superbus		NSW Vul
AMPHIBIANS, REPTILES			
Green-thighed Frog	Litoria brevipalmata		NSW Vul
Stephens' Banded Snake	Hoplocephalus stephensii		NSW Vul
Pouched Frog	Assa darlingtoni		NSW Vul
Giant Barred Frog	Mixophyes iteratus	Comm Vul	NSW Vul
Sphagnum Frog	Philoria sphagnicola	Comm Vul	NSW Vul

Aniseed Myrtle (Ringwood), Tuckers Nob. This species has been submitted to be considered for listing as endangered

INSECTS			
Southern Pink Underwing Moth	Phyllodes imperialis sthn sbsp	Comm End	
FLORA			
Slender Marsdenia	Marsdenia longiloba	Comm Vul	NSW End
Rusty Plum	Neimeyera whitei		NSW Vul
Red-flowered King of the Fairies	Oberonia titania		NSW Vul
Scrub Turpentine	Rhodamnia rubescens	Comm Crit End	NSW Crit End
Native Guava	Rhodomyrtus psidioides	Comm Crit End	NSW Crit End
Ravine Orchid	Sarcochilus fitzgeraldii	Comm Vul	NSW Vul
Rainforest Cassia	Senna acclinis		NSW End
Milky Silk Pod	Parsonsia dorrigoensis	Comm End	NSW Vul
Red Boppelnut	Hicksbeachia pinnatifolia		NSW Vul
ENDANGERED ECOLOGICAL COMMUNITIES			
Coastal Saltmarsh in the New South Wales North Coast, Sydney Basin and South East Corner Bioregions			EEC
Coastal Swamp Oak (Casuarina glauca) Forest of New South Wales and South East Queensland ecological community		Comm End	
Coastal Swamp Sclerophyll Forest of New South Wales and South East Queensland		Comm End	
Freshwater Wetlands on Coastal Floodplains of the New South Wales North Coast, Sydney Basin and South East Corner Bioregions			EEC
Lowland Rainforest of Subtropical Australia		Comm Crit End	
Lowland Rainforest in the NSW North Coast and Sydney Basin Bioregions			EEC
Subtropical Coastal Floodplain Forest of the New South Wales North Coast Bioregion			EEC
Subtropical eucalypt floodplain forest and woodland		Comm End	
Swamp Oak Floodplain Forest of the New South Wales North Coast, Sydney Basin and South East Corner Bioregions			EEC
Swamp Sclerophyll Forest on Coastal Floodplains of the New South Wales North Coast, Sydney Basin and South East Corner Bioregions			EEC

Die Happy, FoKH reserve

FURTHER READING

Atlas of Living Australia, undated. https://www.ala.org.au.

Boland, D.J., 2006. Forest trees of Australia. Melbourne: CSIRO Publishing.

Cadman, T., Schlagloth, R., Santamaria, F., Morgan, E., Clode, D. and Cadman, S., 2023. Koalas, Climate, Conservation, and the Community: A Case Study of the Proposed Great Koala National Park, New South Wales, Australia. The International Journal of Social Quality, 13(1), pp.25-55.

Cerese, B. 2012. The Eucalypt Forests of Northeast New South Wales: A Preliminary Assessment and Documentation of Their World Heritage Values. Sydney: National Parks Association NSW.

Forestry Corporation of NSW, undated. Open Data Site, https://data-fcnsw.opendata.arcgis.com.

Friends of Kalang Headwaters & affiliated groups, Undated, Headwaters Conservation Proposal A nature reserve for the Upper Kalang, Middle Bellinger and Nambucca Rivers available at: http://friendsofkalangheadwaters.com.au/wp-content/uploads/2020/11/Reserve-Proposal-NOV10-A4-web-sml-1.pdf

Love, Ashley & Sweeney, Oisin F. 2015. A Blueprint for a Comprehensive reserve System for Koalas (Phascolarctos cinereus) on the North Coast of New South Wales. National Parks Association of New South Wales, Sydney.

NSW Government, 2024. SEED The Central Resource for Sharing and Enabling Environmental Data in NSW, https://www.seed.nsw.gov.au/.

Scotts, D. Conserving Koala Populations of the Upper Mid-north Coast: preliminary mapping of populations as a basis for further survey, research and planning (2013)

Williams, J.B., Harden, G. J. and McDonald, W.J.F., 1984. Trees & Shrubs in Rainforests of New South Wales & Southern Queensland. Armidale: University of New England Press.

Datasets and maps associated with this proposal available on request; contact: greaterkoalapark@gmail.com.

See also: https://greaterkoalapark.org.

All errors are the responsibility of the editor. This is a living document; comments and corrections are welcomed.

Made in the USA
Monee, IL
04 November 2026

67858828R00079